Implementing Analytics

Implementing Analytics

A Blueprint for Design, Development, and Adoption

Nauman Sheikh

AMSTERDAM • BOSTON • HEIDELBERG • LONDON
NEW YORK • OXFORD • PARIS • SAN DIEGO
SAN FRANCISCO • SINGAPORE • SYDNEY • TOKYO
Morgan Kaufmann is an imprint of Elsevier

Acquiring Editor: *Andrea Dierna*
Editorial Project Manager: *Heather Scherer*
Project Manager: *Punithavathy Govindaradjane*
Designer: *Russell Purdy*

Morgan Kaufmann is an imprint of Elsevier
225 Wyman Street, Waltham, MA 02451, USA

Library of Congress Cataloging-in-Publication Data
Sheikh, Nauman Mansoor.
  Implementing analytics : a blueprint for design, development, and adoption/Nauman Sheikh.
    pages cm
  Includes bibliographical references and index.
  ISBN 978-0-12-401696-5 (alk. paper)
  1. System analysis. I. Title.
  T57.6.S497 2013
  003—dc23   2013006254

British Library Cataloguing-in-Publication Data
A catalogue record for this book is available from the British Library

For information on all MK publications,
visit our website at *www.mkp.com*

Printed and bound in the United States of America
13 14 15 16 17   10 9 8 7 6 5 4 3 2 1

Contents

Acknowledgments

I would like to dedicate this book to:

My parents: I am thankful for their life-long sacrifice, prayers, and support to ensure I have a better life than what they had.

Sarah: My wife and my pillar of support. I can barely get ready for work without her help; writing a book would've been impossible without her relentless encouragement and effort to provide an environment where I could focus on research and writing.

Sameeha, Abdullah, and Yusuf: My kids for their sacrifice of quality time with dad. They would stand with a book for me to read, a ball to play catch, or a board game, but first check if I was finished writing and never complained if I could not make it.

Jim Rappe and Wayne Eckerson: My fellow professionals who reviewed my work and then encouraged that I should write a book on this topic.

Dr. Fakhar Lodhi: My teacher, long-time mentor, and friend who helped me build a technology-agnostic structure of the entire analytics implementation methodology.

Dr. Sajjad Haider: For his extensive help in researching a wide variety of topics across mathematics, statistics, and artificial intelligence.

Keith Hare, Konrad Kopczynski, and Dr Zamir Iqbal: For reviewing the entire manuscript and providing insightful comments and valuable feedback.

Andrea Dierna: My editor at Morgan Kaufmann who worked patiently with a first-time writer, kept providing valuable feedback, and kept accommodating my missed deadlines.

Author Biography

Nauman Sheikh is a veteran IT professional of 18 years with specialization and focus on data and analytics. His expertise range from data integration and data modeling in operational systems, to multiterabyte data warehousing systems, to analytics driven automated decisioning systems. He has worked in three continents solving data-centric problems in credit, risk, fraud, and customer analytics areas dealing with cultural, technological, and legal challenges surrounding automated decisioning systems. Throughout his career, he has been a firm believer in innovation through simplification to encourage better coordination between technical and business personnel, leading to innovative answers to pressing challenges.

He firmly believes in democratization of analytics and has been working diligently the last few years in building analytics systems using well-known and widely available components. He holds a bachelor's degree in computer science from F.A.S.T Institute of Computer Science, Pakistan and lives in Maryland with his wife and three lovely children.

Introduction

Humans make hundreds of decisions in their personal and professional lives every day. A decision is a response to a specific situation (Reason, 1990), such as where to eat tonight, whether to accept an invitation, if a candidate should be hired, or if a promotional discount will help sales. The more background information available, the better the human mind's cognitive ability to make a good decision from a variety of possibilities. Analytics is all about trying to get a computer system to do the same. This book is about analytics—how data is collected and converted into information, how it transcends into knowledge, how that knowledge is used to make decisions, and how to constantly evaluate and improve those decisions.

This is a practitioner's handbook on how to plan, design, and build analytics solutions to solve business problems. When we go about our daily lives and conduct our day-to-day activities, we produce and consume large amounts of data thanks to the digital age we live in. Historically, data was always produced in nature like weather systems and crop yields, but its storage, processing, analysis, and decision making were all done in the human mind through insights gleaned over years of experience and observation—although this wasn't called "data" in those days. The really smart people did well—they were the wise and experienced who advised and influenced decisions in families, tribes, and kingdoms. They always learned from their own and other's past experiences and extracted insights that were used to make decisions that managed day-to-day activities of households, towns, and governments. Fast forward to modern times, with a proliferation of business conducted through computing, and this acquisition and analysis of data becomes mainstream and enables new ways of looking at data. The evolution of this computing complimented with digital communication exploded the amount of data produced and analyzed, putting the numeric system under severe stress (try counting the zeros in 100s of peta-bytes leading toward zettabytes) (Mearian, 2007).

To put the human society's relation with data and its use in context, Ifrah (2002) quoted:

> Suppose every instrument could by command or anticipation of need execute its function on its own; suppose that spindles could weave of their own accord and plectra strike the strings of zithers by themselves; then craftsmen would have no need of hand-work and masters have no need of slaves.

This quote is from Aristotle who died over 2300 years ago, but was prescient in his analysis of where mankind will be heading. A careful analysis of this statement identifies several themes that attract a lot of research, development, and investment today through the field of analytics. Automation, analysis, and prediction are common themes of using data that Aristotle hints about and we, after 2300 years, are finally starting to realize. This should bring some excitement into analytics and into this book, as our belief in data-driven anticipation and automated decisions is part and parcel of our society's function, and it has been evolving for 2000 years and will continue to evolve. Data, information, and computing are going to remain part and parcel of this evolution for decades to come. Therefore, this is not something that will vanish and be replaced by the next cool thing that shows up in a few years. The techniques and technology will change, but the principle of learning from the past is inherent in our psyche.

In this book, the process of collecting data, learning from it, anticipating scenarios, automating decisions, and tuning and monitoring is labeled an *analytics solution*. It is a very complex implementation challenge, and I have attempted to tackle that challenge in this book through simplification.

Analytics is not a new field of interest. There is evidence of mathematics- and statistics-based techniques being applied to business problems going as far back as 1956 (May, 2009). The proliferation of data from global businesses, web-based commerce, smartphones and smart meters, social media and gaming platforms, and various machine sensors found in automobiles, aircrafts, and construction machinery (also referred as Big Data), combined with advances in computing, storage, and specialized software, has brought new energy and excitement in this field. Typically, analytics solutions are built and run by people with advanced degrees in mathematics and statistics and they use sophisticated software packages like SAS™, SPSS, and MatLab (Mathworks)™ to solve niche problems in economics, finance, and sales and marketing. On the other hand, the last 20 years have seen analytical reporting proliferate across all business functions through the implementation of data warehouses and business intelligence systems. Since data is the greatest competitive asset that an organization has (Redman, 2008) and all parts of the organization use it for reporting and analysis through their data warehouse, why is analytics confined to select specialized areas? This question led to the writing of this book.

The purpose of writing this book revolves around four objectives, and the entire material in the book is designed to achieve these objectives. These also serve as myth-busters for myths surrounding Big Data and Business Analytics regarding technology, expertise, and expense.

Objective 1: Simplification. Since the overall implementation of an analytics solution is actually quite complex, the first step is to simplify the concepts, tools, and techniques and explain how they fit into an analytics puzzle. Part of that simplification is defining analytics as well as various topics, such as predictive modeling, regression, clustering, scoring, ETL (extract, transform, and load), decision strategies, etc. The simplified explanation is enough to make use of these concepts in building analytics solutions and yet have the foundation to attempt larger and more sophisticated solutions.

Objective 2: Commoditization. The entire methodology presented for delivering analytics solutions consistently and repeatedly utilizes commodity technology and human skills. Commodity here refers to tools, technologies, and skills that are not proprietary or, in other words, are not very expensive. The foundation of an analytics solution is built on existing business intelligence and data warehouse programs that are now an essential part of information technology (IT) portfolio. This book will show how existing working components of business intelligence can be leveraged to build analytics solutions. The material also covers the merits of analytics solutions built using proprietary resources that are useful for a very focused and specialized business case usually within one industry. There is a serious case argued toward data mining over established statistical techniques to drive the implementation towards a commoditized solution.

Objective 3: Democratization. This is the most important objective and it is tied into the motivation that resulted in this book. The importance and power of analytics should not be limited to a handful of business cases in financial and marketing space. Its use, application, and adoption can harness value out of any business function, such as procurement, facilities management, human resources, field operations, call centers, project management, etc., most of which are typically cost centers and hardly have the resources to adopt analytics. The methodology presented here will show how simplified and cost-effective deployment of analytics (commodity) enables middle management to improve their KPIs (key performance indicators). The myth of the data scientist (Davenport, 2012) is challenged through this objective, and that role is broken into a functional side coming from business operations and a technical side using commoditized implementation rendering the true data scientist's role, limited to a handful of very specialized areas ensuring Data Scientist is not a prerequisite to getting value from data.

Objective 4: Innovation. Innovation here does not refer to either the creative or artistic aspect of product design or to breakthrough ideas that turn around companies and create new industries like smartphones, social media, shale energy, etc. The perspective of innovation in context of analytics refers to more of a process and culture of innovation that needs to be created in an organization (Drucker, 2002). Like the other three objectives, the idea is to use analytics to innovate within existing business operations and improve their key performance indicators for collective benefit. The myth that you can use analytics solutions to come up with brilliant and game-changing strategies is countered with a simpler alternative. Instead of using analytics in one specialized area to generate a 20% improvement in profitability (not a small feat if possible at all), this book takes the approach of using incremental business process innovation in dozens of functional areas with each contributing 2% to 3% to increased profitability and therefore creating a culture of constant improvement and innovation across all facets of the business.

ORGANIZATION OF BOOK

The book is organized in three parts. Part 1, Chapters 1–3, is more conceptual and technology agnostic—the chapters set up the stage, define the terminology, and explain analytics with a simplistic view. Part 2, Chapters 4–6, deals with actual analytics model design, building, and testing, and then putting the model into production for proactive business decisions. Part 3, Chapters 7–11, contains more specific implementation details dealing with people process and technology. Anyone who has worked with data in spreadsheets or been involved with budgeting or some kind of financial planning, sales volume, or revenue forecasts, will find Parts 1 and 2 very easy and simple to understand, and will be able to follow the content with no need for technical knowledge.

Part 1

Chapter 1 on Defining Analytics first differentiates between current business intelligence and reporting types of activities from what analytics. Then it presents simplified definitions of some very complex techniques and concepts differentiating between mathematics- and statistics-based techniques from data mining. People with experience in quantitative modeling will find the definitions and explanations extremely basic and simplistic; they are welcome to skip to subsequent chapters.

Chapter 2 presents a hierarchy and an evolutionary representation of how data should be used from basic utilization to the highest possible value out of data. This hierarchy is titled *Information Continuum,* and it shows how a traditional operational system and its data utilization is different from data warehouse and its data utilization. And yet how analytics and its data usage is different from both data warehouse and operational systems. There is no skipping of intermediary levels to get to the highest levels of Information Continuum. This will be very useful for organizations and teams to assess their current situation and then see how they would reach the holy grail of automated decisions stepping through the Information Continuum stages.

Chapter 3 on Using Analytics is one of two chapters that contribute toward achieving all the four objectives mentioned earlier. More than a dozen examples from several different industries are presented as problem statements and then the definitions from the first chapter are used to show:

- How these problems can be solved using a repeatable process and commoditized solution employing data mining.
- The common patterns or themes emerging across these varying problems to provide a thought process for finding opportunities that can be solved with analytics.

These patterns are critical in creating a culture where midlevel managers look at their operations and activities and identify problems themselves that can be solved using analytics solutions.

Part 2

Chapters 4–6 cover the specific techniques and concepts that make analytics a powerful tool for business improvement. These chapters cover what quants, forecasters, and predictors have been using for over three decades. Not only the model design, testing and tuning is explained but automated decision strategies on models are also covered in detail along with their governance.

Chapter 4 on models and its variables defines the input variables and then explains their design, evaluation, and testing. These input variables are used to build models (e.g., predictive models and forecasting models). Models are stress-tested, tuned, and replaced and this chapter walks readers through the entire process. No deep technical knowledge of databases, computer science, data mining, or statistics and mathematics is needed to learn from this chapter.

Chapter 5 also addresses all four objectives of this book. This chapter is on decision strategies, which is an essential part of analytics but is not a topic that is widely covered in mainstream analytics material. The analytics model is built using the patterns and insights from historical data, but once it is built, it is supposed to be used within operational activities for proactive actions. If a model predicts that customer ABC is going to cancel his wireless phone subscription in the next three months, what should the wireless firm do? That is what decision strategy is about. Design, implementation, and tuning approaches are provided in great detail to help readers make use of models by utilizing the insights from the models. Business operations managers (mid-level managers particularly) will find this very useful because they can treat the analytics model as a black box that IT will implement for them, but the output from that model is directly tied to a business reaction on that insight. Out-of-the-box thinkers and bold mid-level managers will jump on the idea of constantly building new strategies for specific business scenarios and will develop a habit of constantly designing new and innovative strategies on the output of the same model. Functional and business-oriented professionals will find this chapter to be extremely useful in creating a culture of business innovation. With decision strategies leading to automated business decisions comes the need for an audit and control mechanism.

Chapter 6 on Audit, therefore, discusses audit and control and shows how these controls are designed, implemented, and integrated in the solution.

Part 3

The third part of this book is targeted toward IT practitioners who have been involved with data-centric applications—that is, applications that deal with large and complex data-driven activities like data warehouse systems and analytical applications.

Chapter 7 presents a blueprint for an analytics adoption roadmap pilot project. It explains how to pick a problem, find business champions of the idea, and how to deliver the pilot project using existing infrastructure and tools. It achieves that by demonstrating how data warehouse projects are launched, accepted, and adopted by all areas of a business. Just like all managers want reporting to manage their operations, analytics models and decision strategies should follow the same path of demand and delivery.

Chapter 8 on requirements addresses the "chicken and egg" problem where IT asks what you need built and business keeps asking show me what you can do. This dilemma is true for analytics solutions more than for enterprise resource planning (ERP) or customer relationship management (CRM) solutions since they are based on existing business processes. Analytics deals with anticipation of future scenarios and their responses, therefore, business cannot articulate in sufficient detail what exactly they need. The chapter shows how a problem statement is identified using the foundation laid out in Chapter 3 on Using Analytics and then converts that into a formal requirements solicitation process rather than a requirements gathering process.

Chapter 9 takes a real-world example and then builds the entire analytics solution through all its stages just like any other software development methodology.

Chapter 10 covers the roles, responsibilities, and organizational structure for an analytics team that delivers analytics solutions across an entire organization. This chapter also covers various architecture challenges of building analytics solutions since there are various moving parts dealing with large amounts of data on one side and operational integration for automated decisions on the other.

Chapter 11 is a collection of three independent topics that have a direct impact on analytics solutions and the objectives of this book. The three topics (Big Data, Hadoop, and Cloud) are explained to demystify them and make them accessible to IT practitioners dealing with data-centric systems. Removing confusion, jargon, and marketing buzz from these topics combined with other material in this book will allow IT professionals to put these concepts into their proper place for planning, implementation, and deployment against specific analytics projects.

There is a small section at the end of the book titled "Conclusion" that attempts to show how the objectives for this book laid out here, were addressed throughout the entire book and whether the material was successful in achieving its goal.

AUDIENCE

This book is written for two different sets of readers. IT practitioners working or desirous of working in the Big Data space will find the entire book extremely useful for their knowledge and careers. The first two parts of the book are going to be extremely helpful to midlevel managers, graduate students in business, and other professionals involved in business operations who find data-driven business operations an intriguing proposition.

Concept

CHAPTER 1

Defining Analytics

THE HYPE

Analytics is one of the hot topics on today's technology landscape (also referred as Big Data), although it is somewhat overshadowed by the high-profile social media revolution and perhaps also by the mobile revolution led by Apple Inc., which now includes smartphones, applications, and tablets. Social media, mobile, and tablet revolutions have impacted an individual's life like never before, but analytics is changing the lives of organizations like never before. The explosion of newer data types generated from all sorts of channels and devices makes a strong argument for organizations to make use of that data for valuable insights. With this demand and emergence of cost-effective computing infrastructure to handle massive amounts of data, the environment is ripe for analytics to take off. However, like any technology that becomes a buzz word, the definition becomes more and more confusing with various vendors, consultants, and trade publications taking a shot at defining the new technology (analytics is actually not new but it has been reborn with Big Data; see Chapter 11). It becomes extremely difficult for people intrigued by this topic to sort through the confusing terminology to understand what it is, how it works, and how they can make use of it. This happened with ERP and e-commerce in the mid- to late 1990s and with CRM in the early 2000s. Over time, as the industry matures, consensus emerges on what is the definition and who are the dominant players, respectable trade publications, established thought leaders, and leading vendors.

The data warehousing industry was the first to tackle data as an asset and it also went through a similar hype cycle where terms like OLAP (online analytical processing), decision support, data warehousing, and business intelligence were all used to define some overlapping concepts with blurring boundaries. The term that eventually took hold is *business intelligence* (BI) with the data warehouse becoming a critical central piece. BI included everything—activities like data integration, data quality and cleansing, job scheduling and data management, various types of reporting and analytical products, professionals, delivery and maintenance teams, and users involved with getting value out of a data warehouse. BI as an industry has matured and become well structured with accompanying processes, methodologies,

CONTENTS

training, and certifications, and is now an essential part of IT in all major public or private organizations.

While analytics is an extension of BI in principle and it is natural for existing BI vendors, implementers, and professionals to extend their offering beyond reporting into analytics, it should be separated from a traditional data warehousing and reporting definition. We have lived with that definition of BI for the last 10 years and there is no reason to confuse the landscape. It is important to differentiate between BI in which reporting and data warehousing are incorporated, and analytics in which data mining, statistics, and visualization are used to gain insights into future possibilities. BI answers the question "How did we do?" whereas analytics answers the question "What should we do?" BI answers the question "What has happened?" and analytics answers the question "What can happen?" The temptation to replace BI with analytics as an overarching term to refer to anything related to data is going to be counterproductive. The material presented in this book will provide a detailed step-by-step approach to building analytics solutions and it will use BI as a foundation. Therefore, it is important to keep the two concepts separate and the two terms complimentary.

THE CHALLENGE OF DEFINITION

To help demystify analytics and provide a simplified view of the subject while looking at its significance, wide use, and application, we will look at several definitions:

- *Merriam-Webster* (2012): The method of logical analysis.
- *Oxford Dictionary* (2012): The term was adopted in the late 16th century as a noun denoting the branch of logic dealing with analysis, with specific reference to Aristotle's treatises on logic.
- Eckerson's book (2012), which covers the expertise of leading analytics leaders, defines it as everything involved in turning data into insights into action.

These are very broad definitions and don't really help understand analytics in an implementation and technology context. Instead of defining analytics as a dictionary would do, let's look at some characteristics of analytics that can help simplify the conceptual foundation to understand its various moving parts, allow filtering of the marketing hype, and look for pieces needed for specific solutions regardless of what they are being called.

We will use two different perspectives to lay out the characteristics of analytics: one is related to how business value is achieved and the other regards how it is implemented. But the definition of analytics here will be more of a definition of analytics solutions overall and not necessarily a few tools or techniques. The definition will be broad enough to help readers understand and apply it to their benefit, yet it will have specific boundaries to distinguish from other related technologies and concepts to ensure expectations are met from analytics investment.

Definition 1: Business Value Perspective

The business value perspective looks at data in motion as it is generated through normal conduct of business. For this data, there are three variations of value: the present, the past, and the future, in that exact order. When data is created, referenced, modified, and deleted during the course of normal business activities, it lives in an operational system. The operational system at any given time can tell us where we stand now and what we are doing now. The data at a specific point in time is of relevance to people in day-to-day operations selling merchandise, reviewing applications, counting inventories, assisting customers, etc. The data and the activities are focused on very short durations from a time perspective. This data from "Now" is the first variation of business value—that of *present* activities; over time, it becomes stale and is retained for recordkeeping and starts to build history.

This historical data, usually in a data warehouse, can then be reviewed to understand how a business did in the last month, quarter, or year. This information is relevant to business managers since they can see business performance metrics, such as total sales, monthly counts of new customers, total number of service calls attended, total defects reported, total interruptions in business, total losses, etc. This is typically done through reporting. The analysis from this historical review of information provides managers the tools to understand the performance of their departments. This is the second variation of business value—that of *past* activities. This leads into what the managers should be doing within their departments and business units to improve their performance. The historical data doesn't tell them what to do; it just tells them how they did. Historical data is then run through some advanced statistics and mathematics to figure out what should be done and where the energies should be focused.

This leads to the third variation of the business value—that of *future* activities. Analytics (or Big Data) belong to this variation: What should we be doing? Any tools, technologies, or systems that help with that can qualify to be in the analytics space. This business value perspective of analytics is driven from the usage or outcome of the data rather than the implementation. Figure 1.1 explains this definition.

Therefore, the business perspective of the analytics definition deals with future action and any technology, product, or service that contributes towards the action can qualify to be part of analytics solutions. It can be argued that

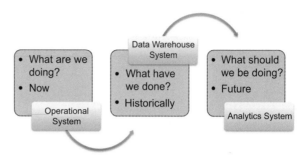

FIGURE 1.1

Analytics definition 1: business value perspective.

historical analysis also gives ideas on what should be done in the future, such as terminate an unprofitable product line or close down a store, but later chapters will show that the analytics solution is actually very specific about future actions based on historical trends and patterns and is not reliant on humans interpreting the historical results and trying to determine the actions rather subjectively.

Another way to define analytics will be from the technical perspective. This perspective has been influenced from the published works of Thornton May on this topic (May, 2009).

Definition 2: Technical Implementation Perspective

The technical implementation perspective describes the characteristics of analytics in terms of the techniques used to implement the analytics solution. If any of the following four data analysis methods are used, the solution is qualified as an analytics solution:

1. Forecasting techniques
2. Descriptive analytics (clustering, association rules, etc.)
3. Predictive analytics (classification, regression, and text mining)
4. Decision optimization techniques

These methods will be defined in more detail subsequently in this chapter, but they will be layman's definitions and will only provide a broad understanding of these methods and their use. For a detailed understanding of these techniques, a lot of good literature is already available in the market like *Data Mining Techniques: For Marketing, Sales, and Customer Relationship Management* by Gordon Linoff and Michael Berry (2011), and *Data Mining: Concepts and Techniques* by Jiawei Han and Micheline Kamber (2011).

If the use of these techniques makes it an analytics solution, why not use this definition alone? Why do we even need the first business perspective definition? Data visualization is what throws off this technical definition. Data visualization deals with representing data in a visual form that is easy to understand and follow. Although it is representing the past, once the visual representation emerges, there are always clusters of data points visually obvious that immediately lead to future actions. For example, in case of geographic information systems (GISs; also clubbed in data visualization), the historical sales data is laid out on a map with each point representing a customer address. Immediately looking at the map with clusters of customers in certain areas, management can decide about advertising in that region, opening a store in that region, or increasing the retail presence through distribution. This geographic cluster of customers is very difficult to detect from traditional reporting. Even though the GIS doesn't use any of the four techniques, it still represents a future actionable perspective of business value.

So why not add visualization to the list of four techniques? It is a subjective argument and open to debate and disagreement. Even though it provides future actionable value, in my view, it is still a reporting mechanism, just more pleasing to the human eye. It is closer to descriptive analytics in definition since it describes the historical data in a visual form, but it is no different than plotting lines, charts, and graphs. Subsequent material (see especially Chapter 5) will illustrate that the implementation of data visualization is very hard to automate and integrate into business operations without a manual intervention looking at the visualization to support automated decisions. Its implementation is also closer to reporting and analytical applications and dashboards rather than, let's say, a prediction or decision optimization solution.

So basically, an analytics solution, technology, or service is only qualified to be labeled as analytics if it fits either of the two preceding definitions. A more conservative way of sketching the boundaries around analytics would be to use both definitions when labeling something as analytics. Text mining is another anomaly that qualifies on the first definition but fails on the second, yet its solution and implementation is closer to prediction even when it doesn't use typical predictive techniques.

The next section explains these selective analytics techniques in a little more detail. The rest of the book relies on these definitions to demonstrate the value of analytics, the design of an analytics solution, the team structure and approach for adopting analytics, and the full closed loop of integration of analytics in day-to-day operations through decision strategies.

ANALYTICS TECHNIQUES

Again, there are four techniques that we will use to define and explain analytics along with its use, implementation, and value:

1. Forecasting techniques
2. Descriptive analytics (primarily clustering)

3. Predictive analytics (primarily classification but also regression and text mining)
4. Decision optimization techniques

Each of these is a science crossing computer science, mathematics, and statistics. An analytics solution will use one or more of these techniques on historical data to gain interesting insights into the patterns available within the data and help convert these patterns into decision strategies for actions. Each of these techniques has a particular use and, depending on the goal, a specific technique has to be used. Once the technique has been identified, the methodology, process, human skill, tools and data preparation, implementation, and testing make up the rest of the solution. The internal workings of these techniques are beyond the scope of this book. There are numerous other books available that explain these in greater detail down to the mathematical equations and statistical formulae and methods. For detailed review of these techniques, use any of the following recommended reading:

- *Artificial Intelligence: A Modern Approach* (Russell & Norvig, 2009)
- *Artificial Intelligence: The Basics* (Warwick, 2011)
- *Data Mining: Practical Machine Learning Tools and Techniques* (Witten, Frank & Hall, 2011)

Algorithm versus Analytics Model

It is important to explain the difference between the algorithm and the analytics model in this context of analytics techniques. Each technique is implemented through an algorithm designed by people with PhDs in statistics, mathematics, or artificial intelligence (AI). These algorithms implement the technique. Then a specific problem feeds data to the algorithm, which works like a black-box to build the model; this is called model training. The analytics model is based on the learning gleaned from the training data set provided. The model, therefore, is specific to a problem and the data set used to build it contains the patterns it has found in the input data. The algorithm is a general-purpose piece of software that doesn't change if the data set is changed, but the model changes with changes in the training data set. Using the same algorithm, let's say a clustering software can be used to build clusters of retail customers at a national store chain or can be used to build clusters of financial transactions coming from third-party financial intermediaries. The model will be different (i.e., the variables and characteristics of a cluster) with changes in input data, but the algorithm would be the same.

An algorithm is like a musical instrument, and the tune you can produce with that instrument is the analytics model.

How good a model you can create given an available algorithm is the same as how good a tune you can create given a particular musical instrument. There is no end to possible improvements in building innovative models, even if using the same algorithm (see Chapter 4 for more explanation on this). The analogy between algorithms and musical instruments further extends: a very high-end and expensive acoustic instrument can certainly improve the sound of the tune you have written compared with an inexpensive instrument. The same is true for algorithms, such as:

- You can get algorithms for free that implement the analytics techniques or ask a PhD student in statistics or AI to build one for you ($).
- You can get algorithms prebuilt in vendor software applications and databases ($$).
- You can buy pure-play high-end algorithms from specialized vendors ($$$).

If you pass the same data set through these three options, some improvement in each progressing model's performance will be evident. But as you cannot create a great sonata by investing in an expensive instrument, an expensive algorithm alone will not result in a quality solution that would deliver business value. Successful application of analytics techniques to business problems requires a lot of other pieces—the algorithm is only one critical part.

The intention of this book is to treat the scientific complexity of a technique and its algorithm as a black-box and explain what they do, what they take as input, what they produce as output, how that output should be used, and where these techniques yield the best value. Readers are not expected to master how to build a new and more robust clustering algorithm, for example, but rather learn how to identify a problem and apply clustering to solve the problem in a methodical and repeatable way. The approach is not very different from a book written about software programming that briefly touches upon the operating system and treats it as a black-box that the programming software interacts with.

Forecasting

Other related terms for forecasting are *time series analysis* or *sequence analysis*. While the people spending their lives defining these terms and carrying out years of research and publishing papers on the finer points of these techniques may be appalled at my clubbing all of these within forecasting, my intention is to simplify the concept so it can be understood and used. Forecasting involves looking at data over time and trying to identify the next value (Figure 1.2). Simply put, if the data points or values are plotted on a graph, then identifying the next value and plotting it on the graph is forecasting. The approach used is very similar to the following problem. Suppose you have a series of numbers like 1, 3, 5, 7, 9, and you are asked what the next number would be—11. How did you derive that value? By looking at the

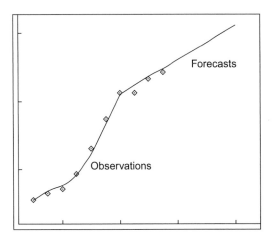

FIGURE 1.2

Forecasting example.

pattern common between each value with its predecessor and successor, the same pattern was applied to the last value to determine the next value. This was a very simple case. Now let's take a little more complex problem. Let's say a series of numbers is 1, 4, 9, 16, 25—what will be the next value? It would be 36. Here the predecessor-to-successor relationship is not as simple as before, but actually a hidden series of 1, 2, 3, 4, 5, 6 is being squared. These examples illustrate the basic principle of identifying a pattern in a series of values to forecast the next value.

Various algorithms exist to provide forecasting like time series analysis, which is extensively used in business for sales and revenue forecasting. It uses a moving average to identify the next value based on how the average of the entire data set changes with each new value. To understand the complexity involved in forecasting, let's look at two extreme examples. Weather forecasting (e.g., what will be the high temperature for tomorrow) can be approached simply by putting down the high daily temperature of the last two years, and based on that, predict what will be the temperature tomorrow. This would be one extreme, that of simplicity, as only one variable, daily_high, is being used and plotted on the graph to see what will be the next value, and it may provide a number that may turn out to be fairly reasonable. However, this is not factoring in barometric pressure, a weather system coming from the west, rain or snowfall, effects of carbon emission, cloudy skies, etc. As we add these additional variables to the model, the forecasting cannot be simply done by plotting one variable on a graph and visually plotting the next value; it would require a more complex multivariable model built from a forecasting algorithm. Time series–based forecasting algorithms are usually embedded in database systems or available as code libraries for various programming environments and can be easily incorporated into an application system. They are used by supplying a learning set to the algorithm so a model can be built. Then the model takes input data and produces output value as a forecast.

Descriptive Analytics

Descriptive analytics deals with several techniques, such as clustering, association rules, outlier detection, etc., and its purpose is to look for a more detailed description of the data in a form that can be interpreted in a structured manner, analyzed, and acted on. If an organization has two million customers, any meaningful analysis of these customers would require questions such as:

- Who are they and how do they interact with the organization?
- What do they buy and how much do they spend?
- When did they last buy something?

Answering these questions for two million customers requires some grouping of these customers based on their answers to get anything useful out of the analysis. A lot of times, in the absence of any descriptive analytics technology, the first 100 records will get picked up for further analysis and actions. Descriptive analytics describes using forms and structures that allow analysts to make sense of mountains of data, especially when they are not sure what they may find. This is also known as undirected analytics where data mining algorithms find patterns, clusters, and associations that otherwise may not have been known or even sought.

For further explanation we will use clustering to explain descriptive analytics in a little more detail.

Clustering

Clustering is a method of grouping data points into clusters based on their "likeness" with one another. This method is used widely in customer analytics where customer behavior and demographics are used to see which customers can be grouped together. Marketing campaigns and other sales and service driven promotions are then designed for each cluster that shows similar historical profiles (similar number of purchases, type of products purchased, gender, income, marital status, zip code, etc.) to manage customers consistently. The most popular clustering algorithm is K-means. This algorithm is commercially available in database systems and from third-party companies. It is also available in open source in the R Project (R Foundation, 2012) and through open-source software like WEKA (Group) that is extensively used in academic circles. Figures 1.3 through 1.5 explain how K-means works to build clusters from a collection of data points. Figure 1.3 shows a simple plotting of points in the Cartesian plane.

K-means relies on Euclidean distance to start to see which points are closer to each other and brings them closer to form clusters. The idea is a point is closely related to another point that is closer to it rather than to a point that is farther away.

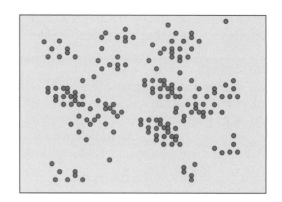

FIGURE 1.3
Plotting of points.

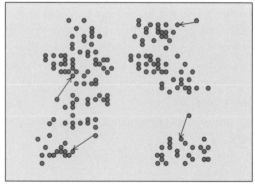

FIGURE 1.4
K-means forming clusters through iterations.

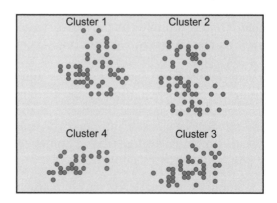

FIGURE 1.5
Clusters formed.

As the algorithm iterates through the plotted points and keeps moving them closer, there comes a time when it can no longer bring the points closer. This is controlled through various parameters that clustering algorithms allow for users to set. At that time, clusters are complete. There can be some points or smaller groups of points not large enough to be clusters that may still lie outside of any clusters; these points are called *outliers* and clustering is used for outlier detection as well as to find anomalies in data.

Once clusters are formed, each cluster has a property that includes aggregate characteristics of the data points in that cluster. So if each data point was a customer, the various characteristics would be average age of the cluster, average income of the cluster, etc. This information is useful for large data sets to break them down into manageable groups so actionable business decisions can be carried out. If you want to give a discount on the next purchase, there is a cost associated with it, therefore it cannot be offered to millions of customers. Clustering helps break down the customer population and then the most appropriate cluster(s) are offered the discount. Clustering is used to understand the population of the cluster by analyzing their common or shared characteristics, but it is also used to see which data points are outliers. Outlier detection using clustering is extensively used in the financial industry to detect money laundering and credit card fraud.

Predictive Analytics

Prediction is the method of predicting the class of a set of observations based on how similar observations were classified in the past (Han & Kamber, 2011). It requires a training set of data with the past observations to learn how certain classes are recognized, and then using that inference on a new set of observations to see which class it closely resembles. This is also known as directed analytics, because through the training data we know what we want predicted. The term *prediction* is used because the model predicts what class the new observation belongs to. Estimation is a specialized prediction where the probability of the prediction is also provided as an output. Classification, prediction, and estimation all have slight variations in their definitions and technical implementations, but for simplicity's sake we will use prediction to refer to all of them.

Let's take a few simple examples. What is the probability that a coupon offer will be redeemed leading to a sale? What is the probability that a given credit card account will result in a default? What is the probability that a manufactured product will have a defect leading to a warranty claim? In each of these cases, historical data or observations with their known outcomes are fed to a prediction algorithm. The algorithm learns from that pattern to see what variables influence the likelihood of a certain outcome (called discriminatory power of a variable), and then builds a predictive model. When a new observation is submitted, the model looks at the degree of similarity of the new observation with the historical pattern to see how the variables of the model (representing the pattern found) compare to the new observation's variables. Then it generates an outcome with a degree of certainty (probability). Higher probability implies the higher likelihood of the outcome and vice versa.

Prediction versus Forecasting

Although during a normal course of conversation the terms *forecasting* and *prediction* can be used interchangeably, there is a distinction between them in the context of analytics solutions as far as training a predictive model is concerned. Within the field of statistics, forecasting deals with time series analysis while regression deals with prediction. This differentiation is important to understand for applying data mining predictive algorithms.

1. The forecasted value could be a new value that may not have occurred in the past. Prediction requires that the historical value must have occurred in the past so it can learn what factors contributed to that occurrence.
2. The forecasted value does not have an associated probability or likelihood of that value occurring (some advanced forecasting models can use probability but it is rare).

Let's use the weather example we discussed earlier. The forecasting problem statement would be: What will be the high temperature tomorrow? Say the answer came out to 72° looking at readings of the preceding days and months. A prediction problem statement would be: What is the likelihood (probability) that it will be 72° tomorrow? If there were days with 72° temperatures in the historical data set, the model will have learned that outcome and identified the variables and values that contribute to it. The model will compare the data available for tomorrow to past days that were 72° and come up with a probability of that temperature occurring. If there was no observation with 72° in the historical data set, then it will simply return 0. There are some advanced forecasting techniques that bridge this divide by introducing probability in the forecasted values, but our emphasis is on simplicity for now and the difference between these two techniques is important to understand for properly using them for specific problems.

Prediction Methods

Predictive algorithms got a big boost from advancements in computing hardware. Faster CPUs, faster disks, huge amounts of RAM, etc., all made data mining–based predictive algorithms easy to design, develop, test, and tune, resulting in widely available prediction algorithms using machine learning. Twenty years ago, prediction problems in healthcare, economics, and the financial sector relied heavily on the regression method for prediction rather than data mining. You will find regression-based predictive models (linear or logistic regression) as the most dominant models used in the financial sector (Steiner, 2012), but regression-based algorithms for analytics are not typically considered data mining or machine learning algorithms even though they have also gotten sophisticated with development in AI theory. The following are popular methods to implement prediction algorithms.

Regression

This is the classic predictive modeling technique widely used in economics, healthcare, and the financial sector. For predicting default in consumer credit space, regression-based models are very reliable and used extensively in risk management within a consumer economy (Bátiz-Lazo, Maixé-Altés & Thomes, 2010). It is through this technique that models were built that labeled certain customer segments as "subprime" because of their high likelihood of default, but lenders decided to ignore those predictions, which led to a global financial crisis in 2008 (Financial Crisis Inquiry Commission, 2011). The regression-based default prediction method was not at fault as far as lending to these borrowers was concerned.

In this particular case, other areas of modeling did fail, and we will cover that in Chapter 5. Regression relies on the correlation of two variables, such as the effect of price increase on demand. They could have a direct correlation, meaning if one increases the other increases as well, or indirect correlation, meaning if one increases the other decreases. In some cases it is possible that no correlation can be observed from the available data set. Price and demand will have an indirect correlation where increase in one will result in decrease of the other within certain constraints. Regression computes and quantifies the degree to which two variables are corelated.

For further reading on regression analysis, some recommended reading materials are:

- Mendenhall, W., & Sincich, T. (2011). *A Second Course in Statistics: Regression Analysis.* Pearson.
- Chatterjee, S., & Hadi, A. S. (2006). *Regression Analysis by Example.* New York: Wiley-Interscience.

In the case of our 72° weather prediction example, the regression algorithm will develop the correlation between the actual temperature and other data variables, such as precipitation, barometric pressure, etc., to build the model.

Data Mining or Machine Learning

The following are three well-known methods within data mining space that can be used to build predictive models. Within the field of artificial intelligence, these methods are also referred as classification methods:

- Decision Trees
- Neural Networks
- Naïve Bayes

age	income	student	credit_rating	buys_computer
<=30	high	no	fair	no
<=30	high	no	excellent	no
31...40	high	no	fair	yes
>40	medium	no	fair	yes
>40	low	yes	fair	yes
>40	low	yes	excellent	no
>40	low	yes	excellent	yes
<=30	medium	no	fair	no
<=30	low	yes	fair	yes
>40	medium	yes	fair	yes
<=30	medium	yes	excellent	yes
31...40	medium	no	excellent	yes
31...40	high	yes	fair	yes
>40	medium	no	excellent	no

FIGURE 1.6
Training data set.

The specifics of their scientific and algorithmic structures are beyond the scope of this book. We will not discuss the merits and demerits of these methods. I have found decision trees to be the simplest to visually represent and explain to people who are new to the field of machine learning or data mining. Therefore, we will use decision trees to illustrate the basic principle of "learning" and "executing"—that is, how a model learns or trains using a data set to create a model, and then how it is used to make predictions on new data sets.

All classification or prediction algorithms predict the occurrence of a particular value of the predicted variable of which the outcome is unknown. Let's take the example from Han (2011) of a propensity model heavily used in customer marketing where some kind of promotion or discount is offered to customers to encourage the purchase of a product. In this case we will use a computer as the product we would like the customers to buy. Figure 1.6 shows the training data, which is basically the historical information on computer sales.

The variables we have represent age, income, whether the customer is a student, and credit rating. The predicted variable is `buys_computer`, and for the training data it has the actual known outcome in the form of yes or no from sales history. The decision tree algorithm will look at the values in the predicted variable and look for the patterns of data in the other variables to build a tree that most reliably predicts the two possible values in this case (i.e., yes or no). The number and quality of the input variable is the key to training well-performing models. Chapter 4 is dedicated to identifying, building, and tuning variables for good models. Out of the four input variables, credit rating is actually a derived or aggregate variable that itself represents complex

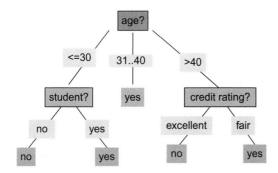

FIGURE 1.7
Trained model.

analytics behind its values. The innovation in creating new variables and their use in models allows for constant improvement to analytics models while keeping the algorithm constant. Identify new variables, tap new data sources, build new aggregate variables, and keep improving the accuracy of the model.

Based on the simplistic data set, the resultant trained model is shown in Figure 1.7. In this figure, the algorithm has identified a pattern that predicts the outcome reliably. This is now a model with age being the variable with the highest discriminatory value to distinguish people who bought a computer and those who didn't. In real-life implementations, millions of records and dozens of variables are used to train models.

With the model now ready, a new record is submitted to this model, and the model will look at the input variables and traverse the nodes of the tree in Figure 1.7 to see where the new record falls. So, if the age on the new record is 23, it will be sent to the left node below age to check the student status. If that is yes, then the prediction result will be yes. So the input data and the model work to predict the outcome when it is not known. This is shown in Figure 1.8.

The basic concept of these examples of clustering and decision trees has been derived from the data mining course materials by Jiawei Han of the University of Illinois at Urbana-Champaign (Han, 2011) and Sajjad Haider of the Institute of Business Administration, Karachi, and the University of Technology, Sydney (Haider, 2012).

Text Mining

Text mining is used to predict lines, sentences, paragraphs, or even documents to belong to a set of categories. Since it predicts the category (of text) based on learning of similar patterns from prior texts, it qualifies to be a predictive analytics method. Although text mining does not use any of the classification or regression techniques, it is conceptually identical to prediction

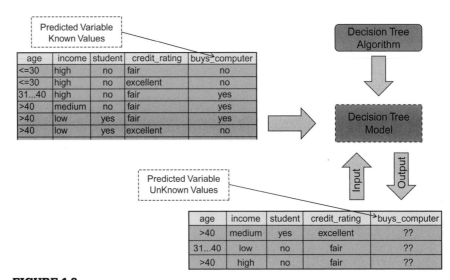

FIGURE 1.8

Predictions or execution.

when it is being used to learn categories of text from a precategorized collection of texts, and then use the trained model to predict new incoming documents, news items, paragraphs, etc. Interestingly, another form of text mining can use clustering to see which news items, tweets, customer complaints, and documents are "similar," so text mining can fall both under descriptive or predictive analytics based on how it is used. To keep things simple, let's take the example of a news story prediction text mining solution. Thousands of documents containing past news stories are assigned categories like business, politics, sports, entertainment, etc. to prepare the training set.

The text mining algorithm uses this training set and learns the words, terms, combination of words, and entire sentences and paragraphs that result in labeling the text to be a certain category. Then, when new text is submitted, it tries to look for the same patterns of terms, words, etc. to see which known category the new text closely resembles and assigns that category to the text. Figure 1.9 illustrates a simple text mining example.

Decision Optimization

Decision optimization is a branch of mathematics that deals with maximizing the output from a large number of input variables that exert their relative influence on the output. It is widely used in economics, game theory, and operations research with some application in mechanics and engineering. Let's use an example to understand where decision optimization fits into an organization's activities. Suppose ABC Trucking is a transportation and logistics company that owns 20 trucks. Trucks make the most money for the

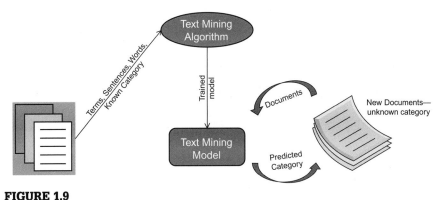

FIGURE 1.9

Text mining.

company when they are full and on the road. ABC Trucking has the following entities or groups of variables:

- Truck (truck type, age, towing capacity, engine size, tonnage, fuel capacity, hazardous material certification, etc.)
- Driver (salary, overtime pay grade, route specialty, license classification, hazardous material training, etc.)
- Shipment order (type, size, weight, invoice amount, delivery schedule, destination, packing restrictions, etc.)

The trucking company wants to maximize the use of its trucks and staff for the highest possible profits, but they have to deal with client schedules, hazardous materials requirements, availability of qualified drivers, cost of delivery, availability of appropriate trucks and trailers, etc. They wouldn't want to travel with the truck half empty, they wouldn't want to pay overtime to the driver, they wouldn't want the truck to come back empty, and they wouldn't want to miss the delivery schedule and face penalties or dissatisfied customers.

The complexity of this problem increases with the incoming shipment orders not having a well-defined or fixed pattern like seasonal volumes. This is the problem that decision optimization models try to solve. It provides the best possible drivers, combination of orders, and routing of the trucks to ensure trucks are not traveling empty, drivers are not overworked, and customers are satisfied. The analysis involves a correlation between variables and maximizing the output. The output in this case would be a routing and shipping schedule that has to be followed during a given time period. This is the scenario where analytics is answering the question of what should be done. Decision optimization always deals with a wide variety of alternates (Murty, 2003) to pick from and tries to find the best possible solution with the given targets. Decision optimization uses linear and nonlinear programming to solve these problems and is usually not considered part of machine learning or data mining. Rather it falls under the field of operations research.

For further reading on decision optimization, some recommended reading materials are:

- Nocedal, J., & Wright, S. (2006). *Numerical Optimization*. New York: Springer.
- Murty, K. G. (2009). *Optimization for Decision Making: Linear and Quadratic Models*. New York: Springer.

Throughout this book, prediction examples will be used to explain the analytics solution implementation methodology and value for solving business problems, but it applies equally to other techniques like forecasting and optimization.

CONCLUSION OF DEFINITION

The most important part of these definitions is to understand the use of the technique so the correct technique is applied to the correct problem. Chapter 3 is fully dedicated to showing a wide variety of problems and applications of appropriate analytics techniques. There are additional analytics techniques such as:

- Market-basket analysis
- Link analysis
- Association rules
- Social network analysis
- Simulation

These techniques all fall under data mining techniques, but are left out deliberately for future discussion as their adoption requires a certain degree of analytics maturity within the organization. Also, the implementation methodology presented here requires some boundaries around analytics techniques and methods, otherwise it would not be possible to show how consistent processes, infrastructure, and manpower can be developed to build and deliver analytics solutions repeatedly. This book is intended to take readers through the introductory stages of adopting analytics as a business and technology lifestyle and become evangelists of this paradigm. Advanced techniques, advanced software, and more complex business problems and solutions will follow naturally.

CHAPTER 2

Information Continuum

Now that we have put some boundaries around defining analytics—a necessary step to provide a step-by-step detailed guideline on conceiving and building analytics projects—we will put some context around the maturity needed to introduce an analytics-driven culture. This context primarily deals with how information is consumed and how that consumption evolves toward a higher value of information use leading to automated decisions. Figure 2.1 is a summarization of Information Continuum—more detail will follow later in this chapter.

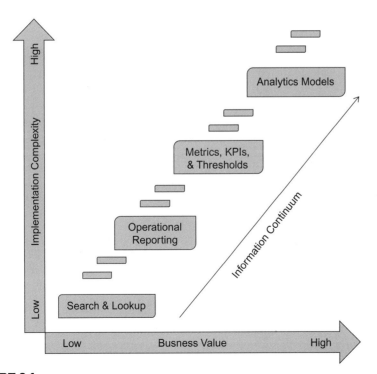

FIGURE 2.1

Introductory Information Continuum.

When the age of computing allowed us to move from paper to data, the power of data—converting it into information, converting information into insights, and making decisions on those insights—dawned on us. It was not because we had no need for information before, but rather the possibilities this paradigm presented. Even in the days of paper, people were converting data into information, but the process was slow, cumbersome, expensive, and error-prone, not to mention severely limiting in the variety and size of data that could be handled. The ability to store years and years of historical data alone catapulted us into a new age of information consumption, making the impossible possible. This chapter presents a hierarchical view of the information need and its utilization over a continuum called the *Information Continuum* (see Figure 2.1). The purpose is to:

1. Show where analytics solutions fit into an evolving process of getting value out of data.
2. Identify the prerequisites before achieving maturity in analytics solutions.
3. Show how technology, people, and processes need to evolve as an organization moves up the Information Continuum.
4. Help organizations assess where they stand today and chart a course toward analytics solutions.
5. Show how business users develop their thought process and trust around data.

Having access to the massive amounts of data does not necessarily mean that the value of information gleaned from that data will be automatic and simple to achieve. The Information Continuum starts at the lowest level of converting data into information for the most basic consumption needs, and then the impact of information use increases considerably as we move up the hierarchy heading toward analytics and decision automation.

The Information Continuum has four building blocks, which require maturity and evolution as we move from the lower levels to the higher levels within the hierarchy. Without maturing of these building blocks organizations cannot create a culture of information-driven decision making. Isolated initiatives can certainly skip some levels in this hierarchy, but the intent is to democratize analytics, and that cannot happen until the preceding levels are achieved along these four building blocks.

BUILDING BLOCKS OF THE INFORMATION CONTINUUM

To yield the most value out of data, the Information Continuum requires four building blocks (Figure 2.2):

- The theoretical foundation in computing, mathematics, and statistics.
- The tools, techniques, and overall technology to help traverse the hierarchy.

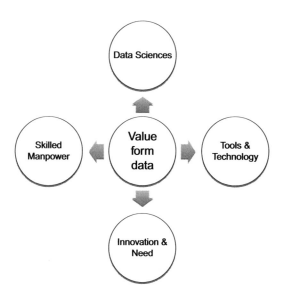

FIGURE 2.2
Building blocks for Information Continuum.

- Trained technical human resources capable of implementing the technology that helps maximize value from data.
- The business innovation appetite to harness the information in new and creative ways.

Theoretical Foundation in Data Sciences

The theoretical foundation around data and information in computer science, mathematics, and statistics has been around for at least 150 years (May, 2009). At the lower levels of the Information Continuum this building block would be arithmetic, algebra, and some basic statistical concepts like mean, mode, frequency, standard deviation, etc., while on the higher end of the Information Continuum would be linear algebra, vector graphics, differential equations, Euclidian geometry, correlation and multivariate regression, neural networks, etc. The explanation of these concepts at the two extremes of getting value out of data is beyond the scope of this book.

THEORETICAL FOUNDATION

Basic to Advanced Concepts Not a Prerequisite

The theory behind getting value out of data includes basic concepts like graphs/charts, averages, mode, frequencies, standard deviation to advanced concepts in linear programming, differential equations, Euclidean geometry, multivariate regression, fuzzy logic, and neural networks, *but* mastery of these concepts is not a prerequisite for designing and building analytics solutions.

Specific concepts will be touched on when they are used to explain some aspect of the analytics implementation methodology. Software development environments or numerous software packages have libraries available that provide these basic to advanced concepts. In essence, the theory exists that can take us from a basic level of information use to higher advanced levels.

Tools, Techniques, and Technology

The tools, techniques, and technology used in extracting value from data at the lower levels of the Information Continuum cannot adequately function at higher levels, therefore, there is a need for the tools and techniques to evolve as well as to deal with newer and more complex information challenges. However, the emphasis is on evolution rather than replacement, so the tools and techniques at the lower levels need to further stabilize and strengthen the foundation for the higher-order tools and techniques to evolve and get more value from data. The biggest example would be the sophistication needed in data management tools, techniques, and technology that deal with much smaller and simpler data sets at the lower levels of the Information Continuum, but are overwhelmed by the size and scale of Big Data at the higher levels on analytics and decision strategies.

Skilled Human Resources

The issue of trained and skilled human resources is critical because availability of the scientific theory and the tools and technology that deliver that theory still require manpower to unlock the potential of data. Data modeling, data integration, data aggregation, and conceptual versus physical structure of the data are the primary skills needed for analytics. This skill is in short supply and will be the Achilles' heel for analytics solutions. The lack of talent in the art of data in enterprises can be linked to the rise of the enterprise resource planning (ERP) software packages and a general shift toward buying products rather than building them.

The enterprise and general maintenance staff was not going to build complex data structures based on their business problems, rather the software vendor was going to address it. Therefore, the skill is concentrated within development teams of large product suite vendors. Data warehousing did give rise to this skill but it is limited to midlevel tiers of the Information Continuum. The material in subsequent chapters will show how to address that gap by helping transition the data warehousing skill set into a specialized data architecture skill set essential for analytics solutions. The perception that pointing analytics software to a large data set will yield results will be put to the test and highlight the importance of data architecture as a critical skill needed to attain higher levels in the Information Continuum with consistent and repeatable delivery of analytics solutions.

Innovation and Need

The business need is central to the success of an analytics solution, because without a clear direction, analyzing data aimlessly can be a never-ending exercise and can result in projects getting shelved and teams getting frustrated. However, in the business personnel's defense, since they do not fully understand or appreciate the power or possibilities of analytics, they may not be able to articulate what they are looking for. The trend is to usually look for a successful project in their industry from conferences, industry publications, or case studies, and then replicate the same with their own data. So the requirements or need may come from competitors, academia, industry forums, etc.

One of the goals of Chapter 3 on using analytics is to democratize the analytical need and show mid-to-low level managers how they can embrace analytics to improve the performance of their departments through a varied collection of detailed examples. And since the entire implementation detail in this book tries to simplify the implementation cost and complexity of an analytics solution, hopefully lowering of the cost and knowledge barrier would encourage innovation and the need for higher analytics. Yet, this is a chicken-and-egg problem where technology will ask what business needs and the business will ask what's possible. A collaborative pilot or prototype using existing data and software is the key to overcoming this hurdle and IT will have to drive toward this.

All four building blocks have to come together and evolve along the Information Continuum to help us move from a very basic conversion of data to information leading into very advanced information value creation.

VALUE FROM DATA

Building Blocks
- Theoretical foundation
- Tools, techniques, and technology
- Skilled human resources
- Information need

INFORMATION CONTINUUM LEVELS

To understand analytics, its capabilities, its use, and its implementation challenges, it is important to understand the Information Continuum because analytics solutions cannot be performed in a vacuum. The prerequisite levels in the Information Continuum have to be achieved first before attempting to democratize analytics for value across an enterprise. Figure 2.3 depicts the

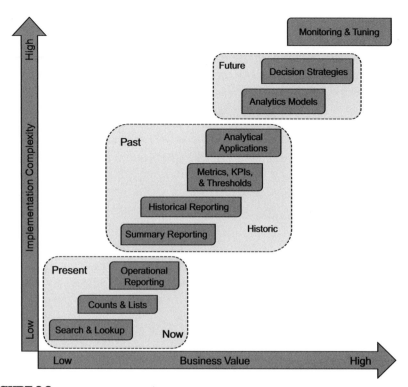

FIGURE 2.3
Information Continuum.

Information Continuum starting with a basic search on data generated from a wide variety of internal and external sources, leading all the way to automated decisions based on analytics, their monitoring, and their tuning.

As more technical and implementation details are offered later in the book, the Information Continuum shown in Figure 2.3 will act as a base to demonstrate what happens at lower levels and how to transform that toward higher levels. This is a continuum, meaning the distinctions between neighboring levels are blurred but become significant as the distance between them increases. In line with the business definition in Chapter 1 where a time perspective of data analysis was used to show how analytics was different from existing methods of getting value from data, the Information Continuum further elaborates that point.

Search and Lookup

The first and most basic level of information need is looking for a specific record. From the entire customer base, a help desk user can look up a customer's account detail while the customer is on the phone. The search mechanism is usually predefined and the user cannot search using any random criteria. A predefined set of fields is typically set up on which a search can be performed. Search and lookup can be for a specific record if the search criterion is detailed enough, otherwise a closely matching list is presented to the user who then selects the record under consideration. The specific traits of this lowest level of information need are as follows.

Implementation

The implementation is built within the transactional or operational system where the raw data is generated and housed. The available technology is sufficiently mature to meet the needs of this level and the manpower to build this capability is also abundantly available.

Challenges

The challenges in this first level of the Information Continuum are driven from two factors, master data and history. If the master data (master lists of all customers, products, accounts, locations, etc.) is not in one place in a standardized format, the search feature becomes quite cumbersome to implement, as it would require searching across numerous systems housing that data. This scenario is typical in merger and acquisition activity where it may take years to identify and assign unique account numbers or identify a search criterion for uniqueness across various business divisions or geographies.

The challenge of history comes from data retention requirements for a search. If a record being searched is eight years old, for example, it may be no longer available as an online search and some manual back-office activity may have to be undertaken to search for that record in archives. Archival and recovery systems are getting better at this and most industries have a regulatory mandate for data retention. If the master data or unique search criterion and historical data requirements are added to this stage of information need, the complexity of technology and cost of implementation goes up, but the business value may not necessarily increase.

Counts and Lists

The next level of the Information Continuum deals with counts or summaries delivered through some form of reporting. The counting could be of transactions in a month, transactions by a customer, or counting of the customers themselves based on some criteria like recent activity, age, or status. Other examples include hours worked by employees in a location, products sold in a season, or simply the sales volume in a quarter. The unique feature of this level is the object or entity being counted, such as sales transactions, products, accounts, hours, volumes, etc., and the selection criteria by which the objects are being counted, such as time period, location, demographics, codes, types, etc. If the information consumer clearly articulates and explains the possibilities of various selection criteria, the resulting implementation is easy to manage, but in the absence of a clear understanding of the different ways in which something can be counted, this can become quite challenging to implement.

It is important to see the transition from a single record-based search and lookup in level 1 to the aggregate counting of those records in level 2. This shows how the Information Continuum demonstrates the evolution of information consumption—that is, from one record search to a collection of records and lists.

Implementation

The implementation mix is both a transactional system and a separate data warehousing type of system. Both options are viable and technology and trained manpower is available to carry out either implementation. The business intelligence tools and techniques have actually matured to address the dynamic selection or grouping criteria, and techniques developed in the data warehousing space are now widely used even within the operational systems to address this requirement. The key is to ensure the definition of the counts and the selection or grouping criteria is well documented and understood by the users of that information.

Challenges

The challenges in this level of information consumption are around the dynamic nature of selection for counting a certain entity. If the ways a certain entity is counted and the ensuing count is used for business purposes are well documented and well structured, the implementation is simpler. However, if the business truly wants a very dynamic mechanism to count customers, sales, or hours worked, then the complexity of providing that capability increases. The data warehousing concept of star schemas or dimensional modeling used with an advanced reporting tool may remain the only option in such a case.

Operational Reporting

This is the most common form of information consumption that has existed long before data warehousing, business intelligence, or even computing took control of business activities. The operational reporting refers to long-established information sets, standardized terminology, and industry- and regulatory-mandated data. Executives and middle management are used to asking for these reports in their regular staff meetings and they are shared with a larger audience across the enterprise. These reports sometimes act as a starting point for the day's activities, such as delinquent tasks from the previous night that need to be picked up, open work orders that haven't been shipped, putting cash in a vault for a bank's branch before opening, etc. Regulatory reporting also falls in this category. Operational reports always have a well-structured format and run frequency (daily, weekly, monthly, etc.).

Implementation

The implementation of this type of standardized reporting has been the driving force behind the data warehousing industry. Prior to modern data warehousing (Inmon, 1992), there were two chronic problems that evaded the wider adoption of reporting in decision making. One was the fact that various systems had the same or overlapping data and, depending on which system was used to run the report, there would be discrepancies (this is commonly known as the problem of "one version of truth"). The other was that reporting consumed large amounts of hardware resources, impacting the operational activities. The modern data warehousing eliminates a lot of these pains and therefore this level of Information Continuum has matured the most.

Challenges

The data warehousing industry has been maturing over the last 20 years. Most of the challenges related to performance, data integration, data quality, and visual presentation of reporting have been adequately addressed, but data warehousing is under severe pressure from shrinking nightly batch windows, explosion in data volumes, and increased sophistication in evolving business needs. The data warehouse was conceived to address a specific set of problems, and expecting it to solve all the challenges coming from higher levels of the Information Continuum is an unreasonable expectation. It remains a central piece in analytics solutions but acts as a source of reliable factory that can deliver data into analytics solutions.

These three levels within the Information Continuum covered so far all deal with the present and are focused on information use concerned with what is happening now. They help answer the question of how we are doing now. There is a subtle maturity evolving across these three levels in information consumption where business went from one record, to lists and counts, and on to formally structured reports. This is the inherent nature of information consumption where maturity at one level automatically leads to the need for the next level.

Summary Reporting

The summary and aggregate reporting is very closely linked to the operational reporting but deals more with historical data than current data. Since this deals with historical data, usually there is a requirement to summarize or aggregate the data before presenting it. The summarization can be from various perspectives like geography or product, but usually the time perspective of summary is used and data is summarized over a time period (monthly, yearly, etc.). Ad-hoc reporting capability is also introduced at this level where users can change the filters and displayed columns on a report.

Implementation

Implementation of this level of information has also become mainstream with the advent of modern data warehousing. Reporting tools (BI tools) deployed in data warehousing systems can easily build this level of information. The tools, technology, human skills, and methodology are all well documented and readily available. The implementation can also be done in an operational system, and it is becoming quite common for operational system vendors to add summary and aggregate reports onto the operational databases through a smaller version of a data warehouse called a datamart that is usually limited to a specific business function. The ad-hoc nature of implementation has to be properly managed so users have a little more flexibility in managing the content on the report through filters and adding additional columns, without getting a report developer involved from IT.

Challenges

Similar to the challenges discussed in the counts and summaries level, the challenges surrounding the summary and aggregate reports deal with how the users want their data summarized. If implemented using a mainstream reporting tool, the users' ability to aggregate the information across any perspective, such as geographies, time periods, etc., improves considerably. If report developers would put the code for the report directly into the tool to run the report, it limits the users' ability to change the summarization parameter on-the-fly. However, getting the tool to dynamically generate the report based on whatever criteria users' input, requires a good design that is based on sound analysis of how the summaries are actually viewed and used for decision making. Therefore, this information level challenge can only be adequately met if all aspects of modern data warehousing are carried out efficiently (Kimball, Ross, Thornthwaite, Mundy & Becker, 2008).

Historical (Snapshot) Reporting

Snapshot reporting is a type of report that allows for comparison of data from two different time horizons. Comparison of a region's quarterly sales with the same quarter last year, profitability, and cost comparison month over month, and activities year over year, such as items produced, calls received, orders fulfilled, etc., are all examples where historical comparisons are made. The historical snapshot or comparison can be between two periods or more. A simple way to understand this is by looking at a summary or aggregate report run for separate periods and then combined in to one report. The snapshot reports are always looking at summarized data.

Implementation

The implementation requires careful snapshotting of historical data so new data does not overwrite the previous information, and the state of the information is frozen at a point in time no matter when you actually view it. This implementation is very difficult to achieve in an operational system and therefore it is recommended to always implement using a data warehouse system and data warehouse design techniques invented for ease of analysis across different time periods (Kimball, 2002).

Challenges

The challenge lies in the design of the data structure to handle historical data frozen in time that can be searched, reviewed, and compared. Several design techniques for this type of data structure exist in widely available dimensional modeling literature (Kimball, 2002). A good balance between what should be snapshotted and tracked over time versus keeping everything snapshotted in time has to be maintained, as data volume, growth, and unnecessary performance pressures may lead to higher cost of maintenance. The reason for this usually is the inability of the business to explain exactly what type of information they would want to compare—that is, they want to keep their options open. In the absence of that analysis, either a minimum level of information is structured for historical comparison or everything is tracked. Tracking everything becomes too expensive and tracking too little becomes a rebuilding pain when new requirements come along. If certain historical information is not tracked, then it cannot be reproduced easily. For example, a large department store's sales volume is stored per quarter and can be compared across several years, but a requirement comes along asking for the sales to be broken up by departments. It is extremely difficult and almost impossible in some cases to break down the quarterly sales data by departments for already closed quarters going back several years.

Metrics, KPIs, and Thresholds

The summary reporting, aggregate reporting, and historical snapshot reporting levels lead into this level of the Information Continuum where various different metrics and measures are developed that business would use to gauge its performance. Selection of metrics and their use is critical to the survival and growth of any business (Kaplan & Norton, 1992). With any metric, to make the information meaningful, there has to be some expected or established pattern or trend. Once a metric is identified, its trend has to be established, and then determine what level of deviation from the trend will prompt a business reaction. This is called setting up a threshold. The metric is reviewed regularly and if the review determines that the metric is breaking the threshold, then a decision is regimented.

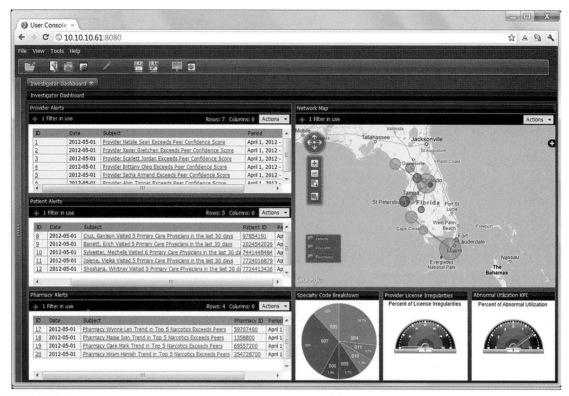

FIGURE 2.4
Dashboard example.

Metrics are the culmination of the summary and historical reporting levels because they condense that wealth of information into a numeric value that can be easily monitored, understood, and tracked. You can use reporting or you can use metrics, but without going through the evolution of the Information Continuum and its preceding levels, it is not easy to conceive a new metric that can provide ample visibility into a complex business operation. Figure 2.4 shows an example of a dashboard (Pondera Consulting, 2012) with actionable alerts on the left side.

Implementation

The implementation of metrics, its trends, and its thresholds is part of business intelligence through an area called dashboards (Eckerson, 2010). Most reporting tools now have dashboards that allow for defining, recording, trending, and visually representing the thresholds. Development in visually appealing tools and techniques has made the dashboards very attractive for business consumption. The real-time updates to dashboards so KPIs are reported in real time to business executives allow for quick reaction to a developing situation.

Challenges

One of the challenges is to help derive the business to a new metric or getting to a point where new metrics are constantly being developed with changing business environments to get a more granular view of business performance. Another challenge deals with the adoption of newer metrics in decision making, as everyone has to perceive the metric exactly as its implementation intended.

Analytical Applications

Analytical applications is where all of the previous levels of the Information Continuum come together. A very popular implementation of analytical applications is the dashboard with attractive displays, colors, and animation. What a dashboard does is basically simplify the information from several summary reports, several historical comparative snapshots, and several metrics, and presents that simplified view to management for fast and effective decision making. This type of analytical application includes triggers and alerts, appropriate email messages, and an overall governance of how metrics are created and adopted.

Another type of analytical application includes data visualization where data is represented through geo-spatial maps, 3D graphs, scatter plots, etc., to detect clusters, spikes, dips, and anomalies. In geo-spatial analysis, data is centered on addresses that are converted into geo coordinates (latitudes/longitudes) and then displayed on a geographic map. The points on the map can be filtered based on any number of factors and uses of historical data, summaries, and even metrics. The important feature of this level is the visual appeal or method of representing data.

Additional interesting analytical visualization techniques will keep appearing in broader markets or very specialized areas, but their nature will place them into this level of Information Continuum. Some degree of statistics and mathematics are also part of analytical applications and, for a predefined business problem, packaged analytical applications are also available, in direct marketing, sales force management, geo-spatial analysis of ATMs and branches in banking, etc. Figure 2.5 is from a fraud detection software package (Pondera Consulting, 2012) where potential collusion between suspects is presented as an overlay of circles representing contacts on a geo-spatial map.

From the summary reporting level to the analytical applications level within the Information Continuum, the focus has been to understand historical data (i.e., the past), learn from it, and then make business decisions. It focuses on how we did as opposed to the first three levels of how we are presently

FIGURE 2.5

Geo-spatial analysis example.

doing. The sophistication of information use continues to evolve across these levels and as the horizontal axis in the Information Continuum (Figure 2.3) depicts, the business value increases as these levels are traversed.

Implementation

Typical implementations of analytical applications require a separate application instance with its own database that follows the requisite structural requirements of the application. Most analytical applications operate in a read/write fashion where user interacting with the system also generate data that is written back. The implementation architecture should have a clear integration interface with the data warehouse which would feed data into the application database. The analytical application is like an operational system as far as user interaction is concerned but they are always downstream from the data warehouse and are fed from it.

Challenges

While the visual aspect is pleasing, this Information Continuum level has to be built once the underlying levels have a wider adoption and appreciation.

The biggest challenge is around interpretation of the visualization. Various departments and groups have different perspectives and ensuring the same visualization is interpreted the same way across the enterprise may require considerable user training and education. If a vendor-driven pilot is undertaken on a specific problem without going through the evolution of the Information Continuum, the results may be pleasing, but the tool's ability to benefit the entire organization may not be achieved. Therefore, the requirements for an analytical application are critical, since identifying and agreeing on the information to be made available through the application is a complicated mix of business needs and technical capability. The entire set of available data is overwhelming and cannot be effectively presented in these visual formats (dashboards, 3D graphs, and geo-spatial). Therefore, the Information Continuum hierarchy has to be followed as along its levels comes knowledge and understanding to interpret the data.

Analytics Models

This book is about analytics and this section will only touch briefly on the Information Continuum level of analytics since we are going to deal with it piece by piece in subsequent chapters. The purpose of this topic here is to put analytics in a context of information evolution, technology, and processes. Analytics solutions today are built disconnected from the rest of the Information Continuum, but until a solution is linked back into the Information Continuum and its overall implementation roadmap, value from analytics would not become commoditized and democratized. At the beginning of the chapter we argued why analytics should be available to everyone in an organization for improved customer experience, operational efficiencies, and increased innovation, across all business functions. If data is available, then standard reporting or summary reporting should be available based on the Information Continuum levels. If summary reporting is available, then historical snapshot reporting, metrics, and analytical applications should be introduced at every management level as well. Therefore, the argument continues that analytics should be available for everyone as well.

As per the definition of analytics, there are four types of techniques involving some combination of data mining, mathematics, statistics, and computer science to deliver analytics:

- Forecasting methods
- Descriptive methods (clustering, association rules, etc.)
- Predictive methods (classification, regression, and text mining)
- Decision optimization methods

We will only focus on descriptive and predictive methods for most of the rest of the book to illustrate how to identify opportunities or problem areas where these methods can be applied. Specialized areas within large organizations use some kind of analytics anyway, but the intent is to bring it down to all facets of a business and really democratize it as reporting has become over the last three decades. Just like operational reporting, employees should be relying on analytics to come into work and carry out their day-to-day activities. Throughout the book, we will try to treat four different types of analytics techniques as a black-box and focus on using them for specific input and output.

Implementation

The implementation of analytics is a very tricky matter and requires detailed introspection. If the business problem area where analytics is being applied is a very sophisticated and critical component for business survival, like risk management in a brokerage or trading business, then the implementation should be higher-end as well. This would mean having a few PhDs on staff, a very high-end analytics tool, and an indigenous and tailor-made solution. That type of implementation is very expensive, and making a business case for that requires a lot of commitment from executive sponsors, as well as clear understanding of the expectation and benefits from the implementation. Since we are trying to introduce analytics at every level of an organization, this implementation approach will not work.

The lower-end analytics capability, especially where the results and the benefits are not quantifiably known, would need free open-source analytics tools like R (R Foundation, 2012) or database analytics like Microsoft SQL Server's Analysis Services. This in conjunction with the data warehouse infrastructure already in place, and the detailed methodology presented subsequently in this book will provide a cost-effective implementation roadmap. Once the sophistication level increases and the business sponsors are on board, advanced tools and high-skilled human resources can be employed. The implementation of analytics models has to be within the existing data warehousing infrastructure and manpower.

Challenges

The two-part challenge for adoption of analytics in the Information Continuum is first identifying the problem where analytics can actually help, and then creating an implementation guideline so this can be achieved cost effectively.

Decision Strategies

Once an analytics model is available, what to do with its output is the next level in the Information Continuum. For example, a predictive model can predict the likely losses from a trade but that will always be with a degree of certainty. It may say there is a 77% chance this trade would result in a loss in the next six months. The decision to trade or not now needs to be further analyzed. However, all trade decisions will get some prediction of potential losses along with a probability, and if all of them have to be reviewed, then the analytics model is of little value. A decision strategy addresses this gray area of decision making by optimizing the value of human input. A decision strategy, therefore, would be to decide on cutoffs and take automated decisions. If a trade's loss probability is (let's say) greater than 70%, then the trade should not be approved. If it is less than 25%, then automatically approve the trade and everything in the middle should be reviewed by an expert. Decision strategies require further segmentation and analysis including what-if analysis to determine the cutoffs, while the model is treated as a black-box that assigns some kind of rating, score, percentage, or metric to the output. Analytics without a decision strategy is like a beautifully cut diamond for which no decision has been made whether it will fit in a necklace, ring, or watch. To make use of analytics models, a decision strategy has to be worked out by developing cutoffs and automation of decisions.

The most famous example of decision strategies is in the consumer credit space in the United States where a proprietary analytics model developed by the FICO Corporation (2012) produces a number called the FICO® Score, and then the lending organization uses it in decision strategies such as:

- Approve a credit application or not.
- Whether an expert underwriter needs to review any aspect of the credit application.
- How much interest to be charged.
- Collateral (or down payment) requirement.
- Other fees and terms of the credit.

The current literature in the market on analytics is not emphasizing the decision strategies enough as the next step in the evolution of information consumption; analytics without the decision strategies will be limited to a few decision makers. On the other hand, a wider implementation of decision strategies to support the operations and field staff can convert all workers into knowledge workers. Knowledge workers are not the ones who have all the relevant data, historical and business context, and analytical capability to carry out their jobs driven from insights. Knowledge workers are workers first; they are in the trenches carrying out the day-to-day operations. They are the ones who apply the knowledge and experience of middle managers or subject matter experts in business operations relying on decision strategies. All aspects of the data across the entire Information Continuum has culminated in a decision strategy. Knowledge workers rely on the actions coming out of the decision strategies and carry them out trusting the process and evolution of the Information Continuum.

Implementation

Implementation of decision strategies requires several technical pieces. First is an analytical piece that helps decide cutoffs and ensuing decisions. Second is the decision engine that actually runs every transaction through the strategy and assigns a decision. Third is the integration of that decision within the operational system. Chapter 5 on automated decisions covers this through in-depth design and implementation level detail.

Challenges

The biggest challenge lies in management culture change since they are not used to enforcing decisions down to the day-to-day operations in such an automated form. The hurdle there is the fear of the unknown because what if the cutoff and ensuing decision turn out to be incorrect. How do we ensure the cutoffs are optimized for that operation? Is that automated decision making threatening the middle and first-level management's effectiveness? A cultural shift in automated decision management is far greater than any technological challenges that pale in comparison.

KNOWLEDGE WORKER

A knowledge worker is someone who is altering his or her day-to-day actions based on knowledge from the Information Continuum–driven decisions. Almost like a habit (Duhigg, 2012), the knowledge worker reacts to a business scenario in real time without thinking.

Monitoring and Tuning—Governance

A careful scrutiny of the subprime mortgage crisis reveals that the absence of governance on analytics models and strategies was one of the main culprits that kept the regulators, auditors, shareholders, and even top management in the dark to the extent of the abuse of subprime mortgage policies and business. When borrowers applied for loans, the lenders almost unanimously relied on the FICO Score generated with the credit report coming from one of the credit bureaus (Equifax (2012), Experian (2012), or TransUnion (2012)). This means that an analytics model was used in the decision making and that is a good thing. However, since the FICO Score correctly predicted higher default probability (hence the term *subprime* (Financial Crisis Inquiry Commission, 2011)), it was the decision strategy that used a lower cutoff for approval. The analytics model wasn't at fault, the strategy design was at fault, although the mortgage-backed securities getting triple AAA ratings were the fault of analytics modelers who had not appropriately factored in variables adequately (see Chapter 5 for more on decision automation).

This level of the Information Continuum on audit and control of analytics (analytics governance) actually uses the metadata from the analytics and the decisions and builds a control mechanism to manage decision strategies from going haywire. To understand this better, let's look at the fictional futuristic projection in movies like *Terminator* that show machines getting so smart that they take over the world. These were intelligent machines and they had the capability to make decisions based on data and intelligence (machine learning or data mining), but when their decision strategies started breaking the boundaries that humans never expected, they became masters enslaving humans. If an audit and control layer was in place reviewing every single decision made by the decision strategy of the machines, then they would've known when to interject and adjust the machine's decision automation parameters or metrics, trends, and models on which the decision strategy is built.

Operational activities generate data. We analyze that data through an evolution of reporting capability to better understand what has happened and then we build analytical models to start working toward better understanding what we should do. These suggested actions coming from analytics become actual decisions through a decision strategy, and then this leaves additional newer kinds of operational data that audit and control picks up. Interestingly, with audit and control, we are back in the present, as the decisions are being done in real time during the course of business operations. Therefore, analytics governance uses operational reports, summary reports, compares historical snapshots, and builds its metrics and KPIs to ensure analytics models and decisions strategies continue to perform at an optimal level for business benefit.

The users of this level cannot be easily defined. The risk management teams in financial organizations and internal auditors and controllers are supposed to be the users who should be doing this anyway. Maybe their methodology should change to ensure this. New possible users could be anyone who has a stake in the organization's success and stability, including employees or employee unions, shareholders, external auditors, or regulators. If an organization truly functions as a knowledge organization and follows the Information Continuum, then maybe this level should be inherently available across all functions of the organization rather than one centralized audit function. The business significance of this level is very high because the absence of this level can actually lead to a complete collapse of the organization or significant losses. The rogue trader syndrome, the conniving and conspiring executives, the sheer stupidity of an individual, or the deception of a con artist have all caused damages to the tune of hundreds of billions of dollars in the last five years alone to companies like Bear Stearns, Lehman Brothers, Olympus Corp., MF Global, JPMorgan Chase, AIG, UBS, Barclays Bank, etc., because analytics-driven decisions were not properly governed and circuit-breaking thresholds and alerts on metrics were not in place.

Implementation

The implementation of this level is actually pretty straightforward. It is very similar to traditional data warehousing, only it is metadata data warehousing, as we are analyzing data about the actual data within the organization. This requires converting the information sets, reports, analytical models, decision strategies, and the actual decision transactions all into data and then store this data in a metadata data warehouse that is then accessed using any standard reporting framework to indicate the stability of the business units, specific functions, and the whole organization.

Challenges

The challenge lies in defining the right metrics and their thresholds that should be tracked during the course of automated decisions and alert when thresholds are breached. The technical challenge is no different than the technical challenge in the KPIs, metrics, and thresholds level, as essentially the same action is being repeated.

SUMMARY

The levels of information consumption discussed in this chapter are not merely a categorization of how information is consumed. They actually lay the foundation for the people, processes, technology, users, and data to be managed in a manner specialized for each level. The idea is to build the higher levels of information consumption through an evolutionary approach. Each new level built on top of the previous levels with maturing resources (human, technology, and process) takes on the more challenging levels of the Information Continuum, leading to proactive, insightful, and managed automation of business decisions. In today's climate, the Information Continuum exists as a fragmented collection of tools and expertise. This limits the benefit of analytics skills and technology confined to the group that owns and maintains the specialized solutions.

CHAPTER 3

Using Analytics

Analytics has a degree of mystery surrounding it almost like a magic box that takes in large amounts of data and, voila, business insights jump out. This chapter will demystify analytics by first explaining the specific problems for which analytics can be used using several examples, rather than a generic statement that "analytics finds insights that business can employ." The patterns or insights have to be in a given context—a problem statement—that will show the relevant data that is needed for that context. Any attempts to blindly put an analytics tool on a large data set may or may not deliver results. It can lead to aimlessly wandering in data and yet not getting anywhere. That type of an approach (also known as data discovery) may work in classic academic research and development type of environment, but not for commercial organizations, as the entire initiative may get terminated after a while. It is advised not to undertake analytics project where problem domain or context is not properly established. Examples in this chapter will help.

Based on the Information Continuum where analytics was put in context of other data and information delivery mechanisms, this chapter presents a variety of problems from numerous industries to see where analytics can be applied. The purpose of these examples is twofold:

1. To illustrate the variety of problems that analytics can solve.
2. To illustrate common themes across these problems, allowing you to find similar problems within your area of responsibility.

For each of these problems, the analytics technique used will be identified along with some idea on sample data that is used and the business value of the analytics output. Once you go through these examples, which are presented in a simplistic way not requiring specific industry knowledge, you should start to think along these themes, and will easily find opportunities within your organizations where analytics models can be tried as a pilot to illustrate its value.

Each of the following examples is a legitimate business problem and is presented here in a structured layout. First, the problem statement is described. In some cases multiple examples are taken from one industry. Then the

CONTENTS

analytics model and its selection is discussed as to which technique is appropriate for the problem statement and why. The same part of the example also covers some details of the analytics solution, such as sample data. Lastly, the third section within each example covers how business value is derived from the application of analytics to the specific problem statement. None of these examples comes from formal case studies, and they are very simplistic representations of the industry and the problem.

HEALTHCARE

The healthcare industry deals with patients and providers on one side where disease, diagnosis, and treatment are important, while on the other side it deals with pharmaceutical and healthcare manufacturers trying to solve challenges in disease and life style.

Emergency Room Visit

A patient visits the emergency room (ER), is thoroughly checked, diagnosed, and treated. What is the probability that the patient will be back in ER in the next three months? This is done to track the treatment efficacy of the ER department.

Analytics Solution

A predictive model is needed that will review the detailed records of the patients who returned to the ER within three months of their original visits and the ones who did not return. It will use a predictive variable called ER_Return and set it to 1 for all patients who did return and 0 for all patients who did not return. The data preparation would require historical ER data going back to three to five years. The grain of the record will be the patient visit, meaning each patient visit will have one record. The variables in the record will be age, gender, profession, marital status, diagnosis, procedure 1, procedure 2, date and time, vitals 1–5, previous diagnosis, previous procedure, current medication, last visit to the ER, insurance coverage, etc.

Note that the data preparation is important here because the vital readings, such as blood pressure, temperature, weight, pulse, etc., all have to be built in such a way that one visit gets one record. This is a requirement for the predictive model. Also the predictive modeling tool is not blindly or aimlessly being applied on Big Data, rather the problem statement determines the grain and structure of data.

The predictive model will take the historical data set, look at all the records that have a 1 in the predictive variable, and find some common patterns of variables. It will then look at the records that have the predicted variable as 0 and find some common patterns of variables. Next it will remove the variables that are common in both and identify the variables that stand out in determination of the 1 and 0. This is the power of discrimination, and each variable gets a metric in terms of its power to discriminate 1 versus 0. All of the variables with their discriminatory powers combined become the predictive model.

Once the model is ready from the historical training data set, it will be tested. We will deal with testing in subsequent sections, but basically 90% of the records can be used to train or build the model and then 10% to test the model. The model is run using the 10% data, but the predictive variable is not provided to the model and it is required to assign the 1 or 0 using its knowledge of the variables acquired through the 90% of the data. The model assigns the 1 and 0 to the 10% data, and then the results are compared with the actual outcome that is known but was withheld when the test data was submitted to the model. For simplicity's sake, we will assume that if the model got 70% of the values assigned correctly, the model is in good shape.

Once the model is tested, as a new patient is about to be released from the ER, his or her record is run through the predictive model. The model will assign a 1 or a 0 and return the record. This determination of a 1 or 0 comes with a degree of certainty i.e., a probability. The degree of certainty will be addressed in subsequent chapters in detail. If a 1 is assigned that means the patient will return to the ER in the next three months, and a 0 would mean otherwise. If a 1 is assigned, the decision strategy will kick in and an outpatient clinic nurse is assigned the case depending on the disease and treatment. The nurse would be responsible for following up with the patient regularly to encourage healthy behavior and discipline in dietary and medication schedules. The nurse may also schedule to bring the patient into an outpatient clinic for an examination after one month. All of this ensures the patient does not overburden the health-care system by another visit to the ER if it can be avoided by managed care.

Patients with the Same Disease
A different problem within healthcare is the analysis of a disease to under-stand the common patterns among patients who suffered from the disease. This is used by drug manufacturers and disease control and prevention departments. The problem is to identify common patterns among a large group of patients who suffered from that same disease.

Analytics Solution
In this scenario all the patients under consideration contracted the disease, so we cannot use the predictive variable with the 1 and 0 approach; that is, we are not trying to use two sets of patients: one with the disease and one with-out. This requires a clustering solution. We would build the customer data set for the ones who contracted the disease, including variables like age, gender, economic status, zip code, presence of children, presence of pets, other medi-cal conditions, vitals, etc. The grain of the data prepared for this will be at

the patient level, meaning one record will represent one patient. The clustering software will take this data and create clusters of patients and assign a unique identifier to each cluster (name or label of the cluster). In addition to the name, it will provide the details of the variables and their value ranges within the cluster. For example, cluster 1 has the age from 26 to 34, while cluster 2 has the age from 35 to 47; cluster 1 has 60% male and 40% female, while cluster 3 has 30% male and 70% female.

Ten is a common number of clusters that the clustering software is required to build, but sometimes it cannot find 10 clusters, which means either more variables need to be added or the required number of clusters should be reduced. Clusters where variables have overlapping ranges say cluster 6 has the age range 24–34, would mean cluster 1 and cluster 6 have a strong relationship.

The purpose of building clusters is to break down a problem into manageable chunks so that various different types of approaches can be employed on various clusters. Clustering is most useful when a clear line of attack on a problem is not identifiable. In this scenario, various types of tests and treatments will be applied to different clusters and analyze the impact. Also, if the patient population is quite large, treatment research requires some mechanism to break down the population size to a manageable chunk. Instead of selecting random patients or the first 100 on a sorted list by age, clustering is a better mechanism, as it finds likeness among the population of a cluster and therefore the variation in data is evened out.

CUSTOMER RELATIONSHIP MANAGEMENT

One of the most common and highly quoted applications of analytics is within the CRM space. Direct marketing or target marketing requires analyzing customers and then offering incentives for additional sales, usually coupons like 20% off or buy one get one free. There is a cost associated with these promotions, and therefore it cannot be offered to all customers. There are two distinct problem statements within CRM—that of segmenting the customers into clusters so specific incentives can be offered, and then only offering to those who have a high propensity to use them for additional sales.

Customer Segmentation

Break down the entire customer database into customer segments based on the similarity of their profile.

Analytics Solution

The grain of the data prepared for this problem will be the customer, meaning one record will represent one customer. Sales histories of individual customers have to be structured in such a way that all the prior sales history is captured in a set of variables. Additional variables can be age, gender, presence of children, income, luxury car owner, zip code, number of products purchased in a year, average amount of products purchased in a year, distance from nearest store, online shopper flag, etc. The clustering software will create 10 clusters and provide the specific values and value ranges for the variables used in that cluster. It may turn out that a cluster is found of frequent buyers who live in the same zip code as a big shopping mall. This is a find from the clustering that may not have been evident to the marketing team without clustering. They can now design campaigns in the surrounding areas of malls where there is a store.

The value here is identification of an interesting pattern and then exploiting that pattern for a creative campaign. The direct marketing campaigns have a fixed budget, and if an interesting cluster has a large population set beyond the budgets affordability, more variables may get added to find smaller clusters through an iterative process. It is also possible that the cluster variables and their ranges are not very useful, and therefore more iterations are needed with increased historical data, such as from one year to three years or adding additional variables to the data set. Typically, once good clusters are found that can be successfully exploited, the marketing teams rely on those clusters again and again and may not need additional clusters for seasonal campaigns.

Propensity to Buy

Once a coupon is sent, the direct marketing teams are required to track how many coupons were actually utilized and what was the overall benefit of the campaign. The higher coupon utilization leads to higher success of the campaign. It is, therefore, desirable to send the coupons only to customers who have a higher propensity to use the coupon. The problem statement therefore becomes: What is the probability that the customer will buy the product upon receiving the coupon?

Analytics Solution

This is a predictive modeling problem, as from the entire population that was sent the coupon, some used it and some didn't. Let's assign a 1 to the people who used it and a 0 to the people who didn't. Feed the data into a predictive modeling engine. The grain of the data would be customer and coupon (that got used or went unused) and sample variables would be customer age, demographics, year-to-date purchases, departments and products purchased, coupon date, coupon type, coupon delivery method, etc. The modeling engine will use 90% of the data and look at the records with a 1 and try to find the variables and their common pattern; then it will look for the 0 records and find the variables and common patterns. Next it will combine the two sets of variables and try to determine the variables with high discriminatory power and come up with a predictive model fully trained. The model will get tested using the remaining 10% of the records. Testing and validation will be covered in detail in Chapter 4. The model will take one record at a time and assign a 1 or a 0 and return the predicted value. It can also return a probability in terms of actual percentage. Internally, the model always calculates the probability; it is up to the modeler to set up a cutoff like every record with higher than a 70% probability should be assigned a 1 and 0 otherwise.

The value of the predictive model here is the coupon cost reduction since it is only being sent to customers with a higher probability to buy. If the model returns a 1 or 0, the coupons are simply sent to the customers with a 1. If it turns out that coupon budget is left over or the customers with a 1 are small in number, then the probability threshold would get reduced. This is where decision strategy comes into play. In this scenario, it is better to get the probability as an output from the model and then the strategy will decide what to do based on the available budget. May be the customers with a higher probability to use (let's say greater than 75%) will get a 20% off coupon and customers with a probability to use between 55% and 75% will get a 30% discount coupon.

HUMAN RESOURCE

The use of analytics in HR departments is not as widespread as in some other industries like finance or healthcare, but that needs to change (Davenport, 2010). In the recruitment department, a predictive model can predict the employees potentially leaving and the recruiters can start soliciting more candidates through their staffing provider network. An excellent case study on this topic shows how Xerox Corporation is hiring (Walker, 2012). In the benefits department, a predictive model can predict whether certain benefits will get utilized by employees and ones not likely to be used can be dropped. Similarly, employee satisfaction or feedback surveys also provide interesting insights using clustering to see how employees are similar and then design compensation, benefits, and other policies according to the clusters. This is an open field of investigation and has a lot of room for innovation and growth.

Employee Attrition

What is the probability that a new employee will leave the organization in the first three months of hiring?

Analytics Solution

This is a predictive model problem since every employee who left the organization within three months will get a 1 and every employee who stayed beyond three months will get a 0. The grain of the data prepared for this model will be at the employee level, meaning one employee gets one record. The variables used can be employee personal profile and demographics, educational background, interview process, referring firm or staffer, last job details, reasons for leaving last job, interest level in the new position, hiring manager, hiring department, new compensation, old compensation, etc. It is as much a science as an art form to see what variables can be made available, for example "interest level in the new position" is a tricky abstract concept to convert into a variable with discrete values. Chapter 4 shows how this can be done. Again, 90% of the data will be used to build the model.

The predictive modeling software will look for common patterns of variables with records that have a 1 in the predicted variable and the same process for records with a 0. Next, it will combine the two sets of variables and try to determine the variables with high discriminatory power and come up with a predictive model fully trained. The model will get tested using the remaining 10% of the records.

A critical mass of employee historical data has to be available for this work. An organization with less than 100 employees, for example, may not be able to benefit from this approach. In that scenario, models are built in larger organizations and can be used in smaller organizations provided there are certain similarities in their business models, cultures, and employee profiles.

Once the model is tested and provides reasonably accurate results, all new employees will be run through the model. This can be a scheduled exercise as new employees get hired on a regular basis in large organizations. As the predictive model assigns a probability of leaving to an employee, the HR staffer works with the hiring manager to ensure employee concerns are properly addressed and taken care of, ensuring the employee has a rewarding experience and becomes a valuable contributor to the team.

Resumé Matching

Another interesting HR problem that can be solved with classification is resumé matching. As defined in Chapter 1, text mining is a special type of data mining where a document is run through text mining software to get a prediction as to the class of that document. A text mining model learns from historical documents where a classification is assigned, such as legal, proposal, fiction, etc., and then uses that learning to assign classes to new incoming documents. In case of HR, the problem statement is" What is the probability that this resumé is from a good Java programmer?

Analytics Solution

Going through the troves of documents with assigned classifications, the text mining software learns the patterns common to each class of documents. Then a new document is submitted and the text mining software tries to determine its class. It predicts the class and its probability. Sometimes it would return multiple classes, each with an assigned probability depending on how the software has been configured. For this particular HR problem, existing known resumés in the HR recruitment database should be labeled according to their class, such as business analyst, database developer, etc. Sometimes even multiple classes can be assigned to the same resumé. Once the model is trained, all incoming resumés should be assigned a class and in some cases multiple classes. Next, we take job descriptions and run them through the text mining software so it can learn the job descriptions and their labels. A decision strategy then looks for the highest matching classes between resumé and a job description and returns the results.

In the absence of any standards for job descriptions or resumé styles, this is a fairly simplistic description of the problem, but it illustrates the innovative use of text mining. This approach is superior to the keyword-based matches widely used today. It can be tuned and, depending on the capability of the software matches and degree of matches, can be built with assigned probabilities. Fuzzy matching (that works on "sounds like") can also be used to standardize terminology in the text before a text mining model is trained. This saves considerable time of the recruitment staff to sift through poorly aligned resumés that ended up matching high because of the same keyword use. A preprocessing can be introduced to format and structure the resumés and the job description and separate out sections like education, address and contact, references, summaries, trainings, etc., and then run text mining to classify the similar sections from job ads and resumés.

CONSUMER RISK

Consumer risk is perhaps the area with a very mature adoption of analytics for business decisions within the banking industry. Consumer lending business includes products like auto loans, credit cards, personal loans, and mortgages. As these products are offered to consumers, the expectation by the lending organization is to make money on interest as the consumer pays the loan back. However, if the consumer fails to pay back the loan, it becomes expensive to recover the loan and in some cases a loss is incurred. Analytics is widely used to predict the possibility of a consumer defaulting before a loan is approved. A consumer applies for a loan and the lender carries out some due diligence on the consumer and his or her profile. For consumer risk, regression has been the dominant analytics method that has been used for almost four decades (Bátiz-Lazo et al., 2010). Later developments in data mining provided an alternative to achieving the same goals using data mining. The following example uses data mining as an option to solve the consumer default risk problem.

Borrower Default

What is the probability that a customer will default on a loan in the next 12 months?

Analytics Solution

The data preparation will assign a 1 to all customer accounts that defaulted in the first 12 months after getting the loan, and it will assign a 0 to all customer accounts that did not default. The grain of the data will be at the loan account level, meaning one record will represent one loan account. The variables will be customer personal profile, demographic profile, type of account (credit card, auto loan, etc.), loan disbursed amount, term of loan, missed payments, installment amount, credit history from credit bureau, etc. The predictive modeling software will look for common patterns of variables with records that have a 1 in the predicted variable and the same process for records with a 0 in 90% of the data. Next, it will combine the two sets of variables and try to determine the variables with high discriminatory power and come up with a predictive model fully trained. The model will get tested using the remaining 10% of the records. The output of the model will be a probability of potential risk of default. Since this is a mature industry for using analytics, the probability has been converted into a score, with higher the score, the lower the probability of default. A lot of lending organizations do not invest in the technology and the required human resources to build models; they simply rely on the industry standard FICO (2012) score in the North American consumer lending markets.

In addition to consumer lending, almost a similar process is used in investment-grade securities default prediction. Models built and used by companies like S&P, Moody's, and Fitch use the same concept to assign a risk rating to a security, and buyers or investors of that security rely on the risk rating.

In this particular case when the predictive model starts delivering a score or a probability percentage, the decision strategy kicks in. If the lender has a higher appetite for risk, they will increase their tolerance for risk and will lend to lower-level scores as well. Not only does this allow for lenders to reduce their risk of default, but analytics also allows for other cross-selling opportunities as well. If a loan applicant comes in at a very low risk of default, additional products can be readily sold to that customer and a specific strategy can be designed driven from the predictive model.

INSURANCE

The insurance industry uses analytics for all sorts of insurance products, such as life, property and casualty, healthcare, unemployment, etc. The actuarial scientists employed at insurance companies have been calculating the probability of potential claims going back to the 17th century in life insurance. In 1662 in London, John Graunt showed that there were predictable patterns of longevity and death in a defined group, or cohort of people, despite the uncertainty about the future longevity or mortality of any one individual person (Graunt, 1662). This is very similar to today's clustering and classification approach of data mining tools to solve the same problem. The purpose of predicting the potential for a claim is to determine the premium to be charged for the insured entity or altogether deny the insurance policy application. A customer fills out an application for an insurance policy and until this predictive model is run, the insurance company cannot provide a rate quote for the premium.

Probability of a Claim

Insurance companies make money from premiums paid by their customer base assuming that not all of the customers would make a claim at the same time. The premiums coming in from millions of customers are used to pay off the claims of a smaller number of customers. That is the simplified profitability model of an insurance company. The insurance company wants to know the probability of a potential claim so it can assess the costs and risks associated with the policy and calculate the premium accordingly. So the problem statement would be: What is the probability that a policy will incur a claim within the first three years of issuing the policy?

Analytics Solution

Similar to what we have seen so far, this is a predictive modeling problem where claims issued within the first three years of an insurance policy will be marked with a predictive variable of 1 and the others with a 0. The data preparation would be at the policy level, meaning one record represents one insurance policy, and the data would include policyholder personal profile, policy type, open date, maturity date, premium amount, payout amount, sales representative, commission percentage, etc. Again, 90% of the data will be used to build the model (also known as training the model) and 10% will be used to test and validate the model. The predictive modeling software will look for common patterns of variables with records that have a 1 in the predicted variable and the same process for records with a 0. Next it will combine the two sets of variables and try to determine the variables with high discriminatory power and come up with a predictive model fully trained.

The banking and insurance industries have been using analytics perhaps the longest. However, even they can benefit from the analytics approach presented in this book that relies on open-source or built-in data mining tools to offset some of their human resources costs and ability to work with a very large number of variables, which humans typically find difficult to manage if they were using conventional mathematical (linear algebra) or statistical techniques (regression). In this case, again the degree of claim probability will determine the premium with higher the probability, the higher the premium. The insurance company can build decision strategies where certain ranges of probabilities have a fixed premium and then more analysis is warranted by an underwriter on probabilities that are lower than the threshold but higher than the refusal cutoff.

TELECOMMUNICATION

In the telecommunication space, free minutes and interesting plans for families of all sizes is a challenge that clustering can solve. Cellular telecoms try to understand the usage patterns of their customers so appropriate capacity, sales, and marketing can be planned.

Call Usage Patterns

The cellular companies will build clusters of their customers and their usage patterns so appropriate service, pricing, and packaging of plans is worked out and managed accordingly. The problem statement is to build clusters of customers based on their similarities and provide specific variables and their value ranges in each cluster.

Analytics Solution

The grain of the data will be at the customer level, so one record representing one customer. This means that their calls, SMS messages, data utilization, and bill payment all have to be rolled up into performance variables like monthly minutes used, total minutes used, average call duration, count of unique phone numbers dialed, average billing amount, number of calls during daytime, number of calls during nighttime, along with the customer profile, such as age, income, length of relationship, multiple lines indicator, etc.

The breakdown of customers into clusters is important to understand the customer groups so specific plans can be built for these groups based on their usage patterns determined from the values and ranges of data in variables used for clustering.

HIGHER EDUCATION

Higher education is not known for using analytics for efficiencies and innovation. There is limited use of analytics in the admissions departments where ivy-league schools competing for the top talent would build a propensity model to predict whether a student, when offered an admission, will accept and join the school. They would rather offer admissions to students who are certain to prefer their school over other schools, because if the students granted admission end up going to other universities, that usually leaves a space open since students not getting admitted in the first round may have already taken admissions elsewhere.

Admission and Acceptance

What is the probability that upon granting admission to a student, he or she will accept it?

Analytics Solution

This is again a classification or prediction problem where people rejecting the admission offer will be assigned a 1 and those accepting the admission will be assigned a 0. The grain of the data will be the student application and some of the variables can be student aptitude test scores, essay score, interview score, economic background, ethnicity, personal profile, high school, school distance from home, siblings, financial aid requested, faculty applied in, etc. As before, 90% of the data will be used to train the model and 10% will be used to test and validate the model. The predictive modeling software will look for common patterns of variables with records that have a predicted variable of 1 and the same process for records with a 0. Next, it will combine the two sets of variables and try to determine the variables with high discriminatory power and come up with a predictive model fully trained.

The value of analytics here is that the best available talent interested in the school is actually given the opportunity to pursue their education at the school. A simple cutoff can be built to see what range of probability to accept will become the cutoff point for offering admissions.

MANUFACTURING

In the manufacturing space, forecasting and decision optimization methods are widely used to manage the supply chain. Forecasting is used on the demand side of the product and decision optimization on maximizing the value from the entire supply-chain execution. The use of analytics in manufacturing is destined for hypergrowth (Markillie, 2012) because of several driving forces, such as:

- Three-dimensional printing that is revolutionizing engineering designs.
- Global supply chains with a large choice of suppliers and materials.
- Volatility in commodities and raw material pricing. Procurement and storage decisions are increasingly complex.
- Customization in manufacturing product specifications for better customer experiences. Customization requires adjustments to design, materials, engineering, and manufacturing on a short notice.

While manufacturing will create newer use of analytics to manage this third industrial revolution (Markillie, 2012) within purchasing, pricing, commodity, and raw material trading, we will use a well-defined problem to demonstrate the application of analytics using data mining. The manufacturing process contains engineering where the product is designed; production where the product is built; sales and distribution; and then after-sales service where the customer interaction takes place for product support. Within the customer support function there is warranty and claims management. It is important for a manufacturing organization to understand warranty claims down to the very specific details of the product, its design, its production, its raw material, and other parts and assemblies from suppliers. There is a two-part problem here: one is to predict which product sales will result in a claim, and the other is to understand the common patterns of claims so remedial actions in engineering design and supply-chain operation can be implemented.

Predicting Warranty Claims

What is the probability that the next product off the assembly line would lead to a warranty claim?

Analytics Solution

The solution would require both a predictive model and a clustering model. The predictive model will use 90% of the data set to train the model and then 10% to test it. It will assign a 1 in the predictive variable to records where a warranty claim was paid, and a 0 to all other products sold. The grain for the predictive model is actually quite challenging because it will need product specification, the production schedule, employees who worked the production line, and customer details for those who bought it. The grain would be at the product level but will have a complex data structure. Another challenge here is that for products where a warranty claim was paid, usually there is good detailed data about the defect and about the customer, but where a customer never called for service, the model may not have the pertinent detail on the customer. It is important to have similar data sets available for any kind of modeling, and both the 1 and 0 records should have identical variables and more than 50% should not be blank or null. Variables may have to be left out if they are available for 1 records but not for 0 records.

The predictive modeling software will look for common patterns of variables with records that have a 1 in the predicted variable and the same process for records with a 0 in the data. Next, it will combine the two sets of variables and try to determine the variables with high discriminatory power and come up with a predictive model fully trained.

Once the model is ready, new products being rolled off the line would get a probability of a claim. The products with a higher probability may need to be moved to a red-tag area and investigated further. Value from such a predictive model is not easily achieved, because on an assembly line, thousands of products roll out that are identical in every respect, therefore, a predictive model may not find variables that are different or that have good discriminatory power. This is where art and science come into play, and the list of variables has to be extended by building performance variables or other derived variables using components' manufacturers, batch or lot of the raw material, sourcing methods, etc., to find variables able to distinguish products that were filed for a claim and ones that weren't. Chapter 4 goes in to more detail on performance variables and model development.

Analyzing Warranty Claims

Analyze the warranty claims and build clusters of similar claims to break down the defect problem into manageable chunks.

Analytics Solution

All the claims data will be submitted to clustering software to find similar claims clusters. The grain of the data will be at the claim level, meaning one record per claim, and the variables can include product details, product specifications, customer details, production schedules and line workers' details, sales channel details, and finally the defect details. The software will try to build 10 clusters (by default) and provide the variables and their values in each cluster. If 10 clusters are not found and only 2 or 3 are, then more variables need to be added. Adding variables, creating variables, and deriving performance variables are covered in Chapter 4.

Depending on the properties of the cluster, it may get assigned to a production-line operations analyst or to the product engineering team to further investigate the similar characteristics of the cluster and try to assess the underlying weakness in the manufacturing or engineering process. This is a classic use of clustering where we are not sure what will be found as a common pattern across the thousands or hundreds of thousands of claims.

ENERGY AND UTILITIES

Temperature, load, and pricing models are becoming the life blood of power companies in today's deregulated electric utility and generation market. An electric utility used to have its own power generation, distribution, and billing and customer care capabilities. With the deregulation in the power sector, now power generation is separate from distribution (the utility), and increasingly the customer care and billing is handled by yet another entity called the energy services company. The utility is now only responsible for ensuring the delivery of power to the customer and charging back the energy services

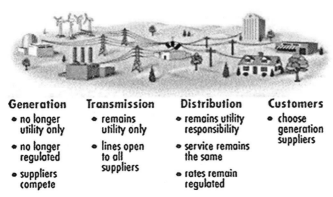

Generation
- no longer utility only
- no longer regulated
- suppliers compete

Transmission
- remains utility only
- lines open to all suppliers

Distribution
- remains utility responsibility
- service remains the same
- rates remain regulated

Customers
- choose generation suppliers

FIGURE 3.1

Deregulated energy markets. *Source: Used with permission of Nexus Energy Software, copyright ©2002 ENERGYguide.com, all rights reserved.*

company that in turn sends a bill to the customer. The energy services company is also supposed to purchase bulk energy from generation companies. The price of energy uses megawatt-hours (MWh) as a unit and it keeps changing based on market conditions and trading activities. A power generation company wants to get the highest price for each kWh that they are producing, and the energy services company wants the price to be the lowest it can be. Usually, a customer pays a fixed rate to the energy services company, but during peak-load periods, the cost of generation goes up and also the price, but energy services companies have to buy and supply energy to the utility so their customers are not without power. This is presented in Figure 3.1.

The New Power Management Challenge

The utility itself is a neutral party in this new environment, but the generation, energy services, and customers all have a significant stake, so analytics will be a critical piece of success in this arena. Turning power off is not an option and that is covered by the regulators. With these two constants (utility and regulators), the other three parties try to tilt the equation in their favor. The power consumption has a consistent behavior in general because factories, commercial buildings, and even households have a very predictable power-consumption pattern. This consistent consumption pattern is disrupted with weather. Extreme heat will force every customer to crank up their air-conditioning and start consuming more power. If the energy services company doesn't have the contracted power capacity to handle the additional load, they may have to buy from generators in the open market. The generators (or other entities holding excess power units) will charge higher tariffs since they may have to tap into additional generation capacity, such as alternate energy sources (wind, solar, etc.), which are typically higher. If the energy services company anticipated this rise in temperature, they may have locked in lower rates already, but now the generator is picking up the additional cost. The customer may also be getting charged a higher rate for peak times as per their agreement with the energy services firm, generating a nice profit.

Analytics Solution

This is a very complex landscape, and it is evolving because the deregulation is starting to take its hold in the marketplace and these scenarios are becoming well understood. Different players in this space will use different analytics methods to maximize their value from this equation. The first thing is the weather forecasting, which of course uses forecasting techniques to see the trending and shifting weather down to the next hour. Organizations may rely on the National Weather Service but usually they need a more granular reading, for example, down to a city block where they may have a couple of large commercial buildings. Next, they have to correlate the weather (in degrees) to load (in MWh). They have to analyze historical data to see the degree of correlation and anticipate the load. Both the power generation and the energy services companies have to do this.

The generation company has to build a predictive model to assess the probability of firing up their expensive power generation plants. They have to manage their costs, raw material, and staffing shifts in line with the probability. The energy services company has to build predictive models on additional factories, commercial clients, and retail customers to come online at the same time as the peak load hits. They have to price their contracts with the generation companies accordingly. They also have to build clustering models to group their customers and their power consumption in "like" groups so pricing plans can be designed and offered tailored to customer needs. This problem is almost identical to telecommunications pricing with peak and off-peak minute utilization plans. Lastly, looking at the load forecast, decision optimization and pricing algorithms have to be used to maximize the profit for energy services companies. The pricing that needs to be optimized is both on the customer side and on the generation side. The generation company has to use decision optimization to factor in oil and coal prices in the global markets along with the cost of plant operation to see what price point works best for them. Both energy services firms and energy generation firms have to now build a negotiation strategy into their pricing mix.

The new energy sector is a wide open space for analytics and its applications. As we realized that multiple models are at work, the decision strategies are the key component that brings the outputs from multiple models and tries to execute an operational decision in near real-time to bid, price, and buy power contracts. Sometimes the excess contracts may have to be sold as power is a commodity that cannot be stored.

FRAUD DETECTION

Fraud comes in wide varieties and we will illustrate two examples that specifically use classification (prediction) and clustering techniques to detect fraud. One example is from the public sector where benefits are provided to citizens and some citizens abuse those benefits. The other example is from banking with credit card fraud.

Benefits Fraud

Governments provide all sorts of benefits to its citizens, from food stamps to subsidized housing to medical insurance and childcare benefits. A benefits program typically has an application process where citizens apply for the benefits, then an eligibility screening where the benefits agency reviews the application and determines eligibility for benefits. Once the eligibility is established the payment process kicks in. The fraud occurs at two levels: (1) the eligibility is genuine but the payment transactions are fraudulent, such as someone having two children but receiving payments for four, or (2) the citizen is not eligible because of higher income level and is fraudulently misrepresenting his or her income. Let's take the problem statement where fraudulent applicants are identified and that information is used to build a predictive model. So what is the probability that this application is fraudulent?

Analytics Solution

This problem is very similar to the loan application and detecting probability of default. This is a benefits eligibility application and we are trying to find the probability of fraud. The solution would work with all known or investigated fraudulent applications assigned a 1 in the predictive variable and a 0 otherwise. Again, 90% of the data will be used to train the predictive model and 10% of the data to validate the model. The grain of the data would be at the application level, so one record is one application. The predictive modeling software will look for common patterns of variables with records that have a predicted variable of 1 and the same process for records with a 0. Next, it will combine the two sets of variables and try to determine the variables with high discriminatory power and come up with a predictive model fully trained.

As the applications come in, they are run through the model and the model assigns a probability of the application being fraudulent. A decision strategy then determines what to do with that probability, whether it is in the acceptable threshold, in the rejection threshold, or if it needs to be referred to a case manager.

Credit Card Fraud

A lot of us may have experienced a declined credit card transaction where the merchant informs the customer that he or she needs to call the credit card company. This happens because the credit card company maintains profiles of all customers and their spending behavior, and whenever they try to do something that violates the expected behavior, the system declines the transaction. This type of fraud detection is also used in anti-money laundering approaches where typical transactional behavior (debits and credits in a bank account) is established and whenever someone violates that behavior, such as depositing an unusually large amount, an alert is generated. The problem statement is to build a "typical" behavior of the credit card customer.

Analytics Solution

This is a clustering problem and the clustering software builds clusters of use for a group of customers based on their behavior. The grain of the data is at the customer level, but all the purchases on the card are aggregated and stored with the customer record. Variables like average spending on gas, shopping, dining out, and travel; full payment or minimum payment; percentage of limit utilized; geography of spending; etc. are built and assigned to the customer's profile. Science and art come into play, as you can be very creative with building interesting performance variables like gender-based purchases, shopping for kids' items, typical driving distances between purchases, etc.

The clustering software will build clusters and assign each customer to its "like" cluster based on the patterns in the data. Once a cluster is identified and assigned to a customer, its variables and their values are established as thresholds in the credit card transaction system. Whenever a transaction comes in breaking a threshold by more than a certain percent, a fraud alert is generated, the transaction is declined, and the account is flagged preventing subsequent use.

The value of analytics comes from preventing fraud from happening, protecting customers, and minimizing credit card fraud losses. This type of a solution cannot function without the integration with the live transaction system and the clusters offer insight into the customer's behavior that can be exploited to offer additional products.

PATTERNS OF PROBLEMS

After going through the preceding examples, readers should start thinking about problems they may be able to attack within their area of responsibility using built-in or open-source data mining tools that allow building solutions economically and efficiently. Each of the problems presented in this chapter is far more complex, and commercial software built by specialized companies may provide a lot more sophistication to solving these problems. Generic models for specific problems are also available from vendors and do not require a training data set, so smaller or newer firms can make use of them.

As these smaller firms mature and have significant data volumes along with enough variety in their data sets, a bespoke model can be trained.

The easiest pattern that can be derived and understood from the preceding examples is the prediction problem. Breaking down the problem into a 1 or 0 allows for predictive modeling software to work very effectively. If any business problem or opportunity can be broken into a 1 or 0 problem (i.e., event occurred or event didn't occur) and data for both sets is available, predictive modeling works really well and is easier to demonstrate business value. Another pattern is tied to the future event's occurrence—that is, if advance knowledge of an event can significantly improve the business' ability to exploit the event, then analytics should be a candidate for solving that problem. However, any problem identified in the business should be worked out all the way to decision strategies, as merely building a model is not going to be enough if it is unaccompanied by innovative ideas on changing the business by adopting the model's output.

How Much Data

The size of data available to perform this type of model building is a challenge. How much data is enough? While there is no harm in having more data—actually the more the better—having insufficient data is a problem. The data insufficiency is in two parts: not having enough records, and not having enough values in variables. For the latter, 50% is a good benchmark—that is, at least 50% of the variables should have a non-null value. But the number of records is a tricky problem. For a rule of thumb, with no scientific basis, for a pilot program on customer propensity predictive modeling, at least 10–15,000 customers' data should be available with at least 12–15% being 1, meaning the event (customer bought in response to a promotion) that we are trying to predict. In case of web traffic data, even tens of millions may not be enough. The data should have enough spread and representation from various population samples and the data set shouldn't be skewed in favor of a particular segment.

Performance or Derived Variables

The number of variables also matters in the quality of a model's performance. Derived variables can be as simple as converting continuous data into discrete data, or deriving a more sophisticated aggregation to come up with new variables. This is covered in more detail in Chapter 4.

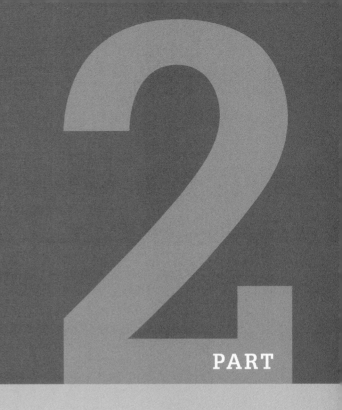

PART

2

Design

Performance Variables and Model Development

Now that we have seen where analytics fits in the Information Continuum and how it can be used to solve a wide variety of business problems, we will move into more specific pieces of input variables, models, and then subsequently the decision strategies. The heart of analytics is the algorithm that takes variables as input and generates a model. The first main section of this chapter will walk you through what variables are, how they are built, how good ones are separated from bad ones, and what role they play in tuning models. The second main section will cover model building, validating, and tuning. The chapter concludes with a discussion of champion–challenger models.

PERFORMANCE VARIABLES

There are a lot of different ways to define variables in an analytics solution. Before we jump into an in-depth discussion about performance variables, Table 4.1 defines the terminology to help you put it in context with

Table 4.1 Terminology for Variables	
Terminology	**Definition**
Columns or data fields	Columns in a database table or fields on a display screen form. Multiple fields make up a record.
Variables	Columns or data fields that become candidates for use in a model and are run through data analysis to see which ones will get selected.
Performance variables	Aggregate or summary variables built from source data fields or columns using additional aggregation business rules. They also go through the data analysis once they are built to select for potential use in the model.
Input variables	These are the variables and performance variables selected for the model as input into the analytics algorithm. This term can refer to both variables and performance variables at times.
Characteristics	Input variables combined with their relative weights and probabilities in the overall model. This term is typically used when explaining or describing a model to business users who will be relying on the outcome of the model to make business decisions.

what is being used in the marketplace by software vendors, publications, and consultants.

What are Performance Variables?

Training records are passed through an analytics algorithm to build a model. Each record is made up of input variables. The model is then deployed and new records are passed so the model can provide a resultant. These records have the same input variables as the training set. Figure 4.1 is the example used in Chapter 1 where we had four input variables—age, income, student, and credit rating—and the predictive model was supposed to learn the relative impact of each input variable onto the prediction about a computer purchase.

The classification algorithm (decision trees) used in this case built a tree with the input variables with highest discriminatory power at the top of the tree and then subsequent variables in branches underneath, leading to a leaf of the tree that has a Yes or No as shown in Figure 4.2.

The data (input variables and historical records) is coming from a sales system that tracks the sales of computers and also records some customer information. There is a limit to how many types of variables are stored in the system. Even if the system is designed to handle a lot of information on a sales record, it is usually not properly filled in. Therefore, there can be a situation where the decision tree cannot be built. There are two possible reasons for this:

1. Not enough historical data records available to learn from.
2. The input variables identified do not have discriminatory power between the outcome values.

The first reason occurs when a sales system doesn't track the lost sales or the customers who didn't buy. If there are no records with the outcome No (the 0 records), it is not possible for the model to be trained for No and it cannot differentiate between Yes and No. Another case could be that the relative proportion of No is so small that it is not enough to build patterns against the dominant Yes (the 1 records) outcomes.

The second reason occurs when input variables are the same values for both 1 and 0 records. For example, 60–70% of the data for both 1 and 0 records of the data set have the same values for age range, or most of the customers have their input variable as Male for gender regardless if they bought or not. In these cases, the algorithm will discard age and gender as input variables because they have low discriminatory value. Here is a detailed example

age	income	student	credit_rating	buys_computer
<=30	high	no	fair	no
<=30	high	no	excellent	no
31...40	high	no	fair	yes
>40	medium	no	fair	yes
>40	low	yes	fair	yes
>40	low	yes	excellent	no
>40	low	yes	excellent	yes
<=30	medium	no	fair	no
<=30	low	yes	fair	yes
>40	medium	yes	fair	yes
<=30	medium	yes	excellent	yes
31...40	medium	no	excellent	yes
31...40	high	yes	fair	yes
>40	medium	no	excellent	no

FIGURE 4.1

Input variables example.

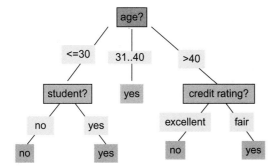

FIGURE 4.2

Trained model.

to explain the reasons why performance variables are essential for a model to perform well.

Reasons for Creating Performance Variables

Suppose a firm ABC Computers sells computers online. They have an order management system that takes orders online, over the phone and fax, as well as through email. The order management system has the data fields shown in Table 4.2 available in its database.

Regardless of how much depth a system has in its design to capture all sorts of relevant data, we know that reality is different, and customers want to provide minimal information while the sales people focus on closing the sale

Table 4.2 Data Fields Example

Entity	Available Data Fields
Customer	Name, address, credit score, age, income, repeat customer, profession, education level, budget, business or personal, marital status, presence of children, promotions, coupons sent, credit rating, etc.
Product	Product ID, product type, class, weight, screen size, processor, memory, hard disk, other specifications, price, lot, manufactured date, Mac, PC, etc.
Sales channel	Channel ID, channel category, channel partner, channel cost, etc.
Sales transaction	Transaction ID, transaction code, timestamp, channel, product, customer, price, payment method, discount, coupon, transaction amount, etc.

instead of recording data. Here is what data analysis will show if we have a few years of sales data recorded in this system:

- A lot of the data fields on various records will be empty.
- If data is filled out, there will be quality issues.
- A lot of data fields will have the same values.

Let's define a rule of thumb as an illustrative data profile of this sales system. Let's assume the total number of fields sourced is 70:

- Total number of fields available: 70
- Fields mostly with null values: 23 (roughly one-third will usually be empty)
- Out of 47, fields with good data: 24 (roughly half of remaining fields will have quality problems)
- Variables available as input variables for the model will be 24 (i.e., one-third of the original field count)

Out of the candidate fields for input variables, two-thirds will be lost to nulls and bad data-quality issues. This is a rule of thumb for guidance purposes only, so start with a larger set of fields across operational systems.

These 24 variables will now be put through a second set of tests to assess their viability as input variables. This is done to make sure the good-quality data filled out has some diversity. So if the field has the same values for all customers who bought and ones who didn't buy, the fields will not be useful. For example, a field with gender has two possible values, M and F, but if all customers who bought and the ones who didn't buy were M, this field is of no use. It is entirely possible to lose another 10 or so fields, leaving us with 14 input variables. Let's assume we have a minimum of two years of data available. If we don't have enough historical data available we have to stop. But how much data is enough? It is very difficult to answer this question generically without understanding the specific business scenario, however; here is another rule of thumb (with no scientific basis) as a guideline or at least a starting point to work with. If the total business transactions activity where the model is being built is 100 in a calendar year, make sure 40 transactions over the last 12 months are available.

For building analytics models, look at the last one year of total business activity and make sure 40% of that is available for model training—35% will be used for training and 5% will be used for testing. This is a rule for the minimum; there is no limit for the maximum—use as much data as available.

Now when we try to build the model by passing the training set through the decision tree algorithm, the algorithm will try to look at combinations of the 14 input variables and their various values, and will try to find patterns that can reliably distinguish between the 1 and 0 records (respectively, yes or no in this case), meaning customers who bought and customers who didn't buy.

What If No Pattern is Reliably Available?

This is very important to understand, because this is the inherent nature of analytics. You may not find a pattern that gives you actionable insight. So the entire exercise of data collection, preparation, cost of hardware and analytics algorithm, and human hours—all of that went to waste? This is entirely possible if proper planning is not undertaken beforehand. You may end up without a model or a model with weak performance (more on that later) to show for all the analytics investment. Performance variables come to the rescue in these types of situations. Performance variables are aggregate variables built using historical and trending data, and they are built at the same level of grain as the remaining input variables. In our example, one record of training data represents one customer who had either bought or had not bought a computer. So if some customers had prior sales transactions in the system, a performance variable will be built at the customer record level to capture the previous sales. Table 4.3 shows some examples of performance variables in the context of our computer sales example.

As you can see from the table, there is no end to the creativity or possibilities of how new performance variables can be built. So let's say the original 14 input variables didn't result in a good model. We add the 5 performance variables from Table 4.3 and now we have additional input variables that may be able to come up with a strong pattern that indicates BUY = Yes and another strong pattern that indicates BUY = No. If these 5 performance variables don't

Table 4.3 Performance Variables for Computer Sales Example

Performance Variable	Data Type	Definition
Purchased_in_last_12_mo (more variables can be created for 6 months, 18 months, etc.)	Char(1)	Set to 1 if there is at least one sales transaction for the same customer in the last 12 months; set to 0 otherwise.
No_of_visits_online	Number	Pull data from the website to see if visits of the customer can be tracked and counted.
No_of_cust_srvc_contacts	Number	Pull data from CRM system to count prior service or sales enquiry calls.
1st_ever_purchase_date	Date	What is oldest purchase transaction date?
Last_purchase_date	Date	What is the last purchase date?

yield good results, go back to the drawing board, analyze the data again using traditional reporting and drilling methods, and see what else can come in handy. Tap into other systems, such as warranty or manufacturing systems, to track the parts and suppliers and build aggregate performance variables using additional data, and run the model again until desired results are achieved.

Benefit of Using Performance Variables

Performance variables are a very important aspect of the analytics solution that can save the investment of an analytics project. Creative use of available data and business knowledge should be employed in constantly finding new and interesting performance variables and then test the performance of the model.

1. Even if desired results are achieved, the performance variables should continue to be designed and models built and tested as an ongoing tuning and model improvement exercise.
2. If prebuilt models are purchased for a specific task, they can be kept from getting stale and irrelevant by adding new performance variables to them.
3. If the software investment included a prebuilt model as well as an analytics algorithm like a neural network or decision tree, then the internal team can keep creating newer and more efficient models and yield long-term benefits from the investment in the software.
4. As business evolves, removing some input variables and adding newer performance variables ensures the model predictions are reflective of changing business dynamics. For example, adding online activity- or social media activity–related performance variables to a customer or product profile can be very helpful to incorporate the impact of those channels.

5. In situations where lines of business or categories of products and customer segments are essential in the business dynamic, it may be worthwhile to get a specialist consulting firm to build a baseline model and then use performance variables to create additional models from that for other lines of business or other customer segments.
6. Performance variables allow for the adoption of industry best practices but then provide an added competitive edge in building superior models above and beyond what the competition has. Data from the same ERP system, for example, can be put through a rigorous and innovative performance variables design activity to add more value into the analytics models.

Creating Performance Variables

While good performance variables can significantly enhance a model's performance, poorly built ones can be detrimental to the analytics exercise overall. It is, therefore, important to carefully consider the following principles while creating new performance variables.

Grain

In the examples in Chapter 3, we emphasized the importance of grain and identified the grain in each example. Grain is the level of detail at which the analytics algorithm will work. An analytics algorithm cannot work at varying grains without some data preparation (ETL) effort bringing it to the same grain. In a clustering example, if we had some points representing customers and some representing households, the intermixed cluster would not make any sense, as the variables used for "likeness" would be different at the household level (e.g., there cannot be one gender representing a household). The grain is tied to the problem statement, and reading the problem statement can explain what the grain is. Similarly, if the problem statement suggests multiple grains, then it needs to be revisited and further refined until one grain is identified.

Range

The range of the performance variables refers to the possible values in a performance variable. Let's look at a performance variable `age_of_position`, which represents the aging of a certain type of security that a trader is holding and it is recorded in days and updated every day. The `age_of_position` can have a range of values from 1 to N. This is called a continuous variable. A variable that has potentially a very large set of possible values and the range of values that may occur on a record is not constrained by any business rule, then it is a continuous variable. The possible values assessment has to come from source system analysis where the data for the variable comes from. If the source system does not put any restrictions on what possible values it can take on, then in theory it can take on any number of unique values, and therefore it is continuous. Continuous variables are not good performance variables and should be avoided if possible. Later in this chapter we will cover how to convert continuous variables into discrete variables.

Spread

The spread of a variable refers to how the population (the entire record set or sample being considered for model creation) is broken out on particular values of the variable. This is the frequency of each value of the performance variable. Let's use an example to explain this.

Suppose we are building a predictive model that predicts the likelihood of an insurance policy to redeem a claim. As part of that predictive model we designed a performance variable representing the customer profitability that stores the amount of money the insurance company has made on that customer since the inception of the policy. The performance variable has the following details:

- Name of the performance variable: CUST_PROFIT_LTD_AMT
- Descriptive name: Customer profitability life-to-date amount
- Data type: Decimal with length 18 and precision 4 (i.e., up to four decimal places)
- Value stored: Dollar figure
- Value range: −$1,200 to $7,300.

The value range is determined by performing the minimum and maximum functions on the performance variable when it has been calculated for all the customers. Now let's say the total number of customers is 10,000 and we have their CUST_PROFIT_LTD_AMT values calculated. This would be a continuous variable because there is no business rule or restraint on what can be the possible profitability of a customer. A continuous variable will usually have a very sparse frequency chart because each individual unique value may have a very small number of customers. Therefore, the frequency would be 1 or 2 for most of the values. This is not a good spread for a performance variable. Similarly, if most of the population is skewed within a handful of values and the other possible values hardly have any population, that is bad as well. Good performance variables should have a good even distribution.

Designing Performance Variables

The design of performance variables uses the following techniques to convert detailed data into useful performance variables. These techniques can be used in any of the four analytics methods we have covered in the book.

Discrete versus Continuous

As explained earlier in this chapter, continuous variables are not good for analytics models to learn from and assign appropriate weights. Therefore,

Table 4.4 From Continuous to Discrete

CUST_PROFIT_LTD_AMT Values (Continuous)	No. of Customers	CUST_PROFIT_LTD_AMT Groups (Discrete)	No. of Customers
$1,200	5	Greater than 1,200	0
$1,159	10	Between 1,000 and 1,200	22
$1,125	7	Between 800 and 999	1
$870	4	Between 600 and 799	1
$730	1	Less than 600	17
$560	9		
$520	8		

continuous variables have to be converted into discrete variables with a finite and small manageable set of potential values. The way to do this for numeric values is to build ranges of values and assign them codes. Table 4.4 provides an example.

The right two columns show a discrete set of ranges that should cover the data set of the left two columns. The design question becomes: How many buckets should be built and what should be their ranges? In this example, we have built five buckets that are 200 points apart. This should not be an arbitrary process of coming up with any number of buckets with any range of values in each. If we draw a bar graph for the two columns on the right, we get what is shown in Figure 4.3. The bar graph should be consistent for the distribution of the population within each range. So the desire is to create a graph like the one shown in Figure 4.4.

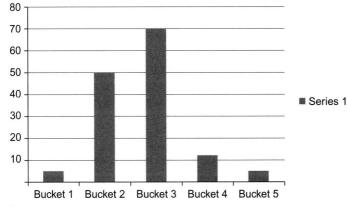

FIGURE 4.3
Buckets for CUST_PROFIT_LTD_AMT.

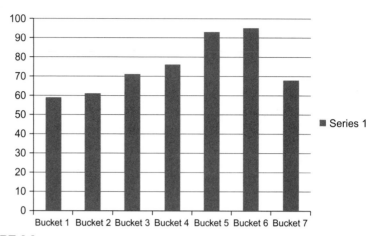

FIGURE 4.4
Preferred Bucket Distribution.

The population is distributed more equally across the values in Figure 4.4. The Frequency function within Excel is a great help in figuring out the value ranges. Each bucket represents a range of continuous values from the original variable. The new variable should be labeled something meaningful to represent the underlying data and its grouping. The buckets can also be coded, meaning a rank or code is assigned to each of them. The ETL process within the analytics datamart will be responsible for building this variable once the ranges have been programmed into the ETL logic. It is entirely possible that in certain situations it is not possible to find such ranges that can evenly distribute the population. In those cases, just use an 80/20 rule to make sure at least 80% of the data to be used in the model training is distributed according to the chart in Figure 4.4.

For other types of continuous data like addresses, try using cities or zip codes to group them in a finite set of manageable values. Strings-based values are almost always continuous, and it is very difficult to convert them into a discrete set of values. The only available option is to convert them into some kind of a higher grouping with a code representing several continuous values.

Nominal versus Ordinal
Another design principle to consider while building performance variables is nominal values versus ordinal values. Ordinal values have an inherent order in the values, so a value of 100 and a value of 200 would imply that the second value is of higher order than the first. Nominal values mean there is no inherent relationship between the values, therefore 100 should be treated the same as 200. When building discrete value sets for continuous data, care should be taken not to create ordinal values or make sure the data mining software has a parameter that can be set to not treat the values as ordinal.

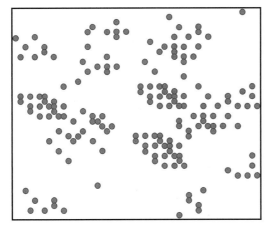

FIGURE 4.5
Scatter graph for customers.

Atomic versus Aggregate

Atomic data or input variables as described earlier in the chapter typically do not fall into the performance variable definition as performance variables are built from atomic variables. They are summaries, intervals, ratios, etc., but are typically not readily available in operational systems.

Working Example

As described in Chapter 3, the art and science of analytics come together at the performance variable level. To understand their impact on the performance of an analytics model we have to be clear on the definition of grain. Analytics algorithms may be K-means algorithms for clustering, regression or neural network algorithms for prediction, or optimization or forecasting algorithms. They work with millions and millions of records at the same grain, meaning all records represent the same entity. For example, one record represents one customer or one record represents one account. They cannot work with one record being a customer and another being an account. So if we are trying to segment our customers into clusters, it would mean that the input data to the algorithm will be one point plotted per customer. Figure 4.5 shows an example of clustering to explain the concept of grain.

Each point represents one customer record with several variables, and the values of those variables have been resolved to result in one point. After plotting them, the clustering algorithm goes through numerous iterations to start to build clusters out of this scatter. The end state looks something like Figure 4.6 where the algorithm puts them into clusters and labels them. In this process one dot represents one customer and therefore it is at the customer grain. If we were using only one variable in the customer record, let's say customer

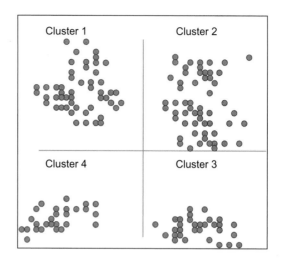

FIGURE 4.6
Clusters.

ID, the algorithm would not be able to find any similarities to bring the dots closer, and the end state of the model would still be the same scatter plot and the tool would probably label it as cluster 1. Therefore, the entire customer population would become one cluster, rendering the effort useless.

For the algorithm to figure out the "likeness" in these dots or points in a multidimensional Euclidean plane, it needs more than just the customer ID to establish likeness. Let's say we add two more variables, age and gender, into the record and try again. If our customer base is more or less in the same age range—for example, a sports drink manufacturer will have the majority of their customers in the 17–37 age range and may be skewed in favor of males at 78% and females at 22%—then the algorithm will still struggle to find multiple clusters. It may find just three to four clusters with each having millions of customers, still rendering it useless for any effective campaigning. Increasing the variables will help ensure the variables are not skewed toward certain values within their range. But adding too many variables may contribute to losing likeness and very small clusters of 10–15 customers may end up clustered with some very large ones. So the variables, their value spread, and their ability to dissect the population are central to getting the clustering algorithm to produce good results. It is important to note that the same clustering algorithm is being used, but the variables and their quality are determining the quality of output from the model.

Using this as the premise, if we are to add interesting variables, since one point can only have one customer, those additional variables have to be at the customer record level. So a variable like Popular_Products_Purchased that can have multiple values per customer cannot be built just like that. It will have to be transposed to act like columns on a customer record and therefore limit it to a handful of popular products (maybe three or five). So the customer record will get three columns: `POP_PROD_PURCHASE_1`, `POP_PROD_PURCHASE_2`, and `POP_PROD_PURCHASE_3`. These are called performance variables since they exist in a different grain of data, but they have to be built to bring them to the grain of the record where the analytics algorithm is being applied. The definition of "popular" has to be agreed on and programmed on the detail-level data and then pick the top three that become performance variables.

Performance variables are needed because they can spread the data in a good mix that allows the algorithms to produce better results. This is why interesting variables can create very powerful models.

MODEL DEVELOPMENT

An analytics model is the heart of an analytics solution, as it is responsible for producing the desired output. The better the model is, the better the value of the output for business decisions. This section covers what a model is, what it looks like or what it contains, how to validate its effectiveness so decisions can be made, and finally how to detect the weakness in a model and tuning it accordingly.

What is a Model?

As described in Chapter 1, it is important to differentiate between a model and the algorithm. Data passes through an analytics algorithm to produce a model. The model takes input records and then produces an output (forecasted, descriptive, predictive, or optimized). In this section we focus on descriptive and predictive models only. Forecasting and optimization models in their description and layout are very similar to predictive models and both techniques have been around for a long time. There is also not much of a mysterious fog on the magical powers of forecasting, since there is nothing hidden within a data mining algorithm or software like we have in predictive and descriptive analytics software. As you will see in the following, forecasting models can be a simple linear algebra equation not requiring the explanation needed of a predictive or descriptive model.

Model and Characteristics in Predictive Modeling

If you want to touch and feel a model, in the following example you can find what it would look like once produced. Some software packages do not allow for easy access to this level of detail on a model. There are good reasons for this, because if input and exact weights of characteristics within the model are known, when a model runs, input can be manipulated to get the desired output.

Sometimes leaving a model as a black-box may not be a bad idea. Following is an example of a consumer risk model that takes input data from a loan applicant and tries to produce a risk score. A high score indicates lower risk and vice versa.

- Characteristic 1: Age
- Input variable: `Age_range_code`
- Data type: String
- Maximum weight: 200

Value	Weight
17 or below	Policy reject, return zero score
18–24	100
25–45	150
46–62	200
63 and above	125

- Characteristic 2: Previous loan status
- Input variable: `12_mth_prev_loan_history_status`
- Data type: String
- Maximum weight: 250

Value	Weight
One or more loans and up-to-date payments	200
No loans before	−50
One loan where two payments have been late in the last 12 months	−100
One loan where one payment was late in the last 12 month but not in the last 6 months	150
More than one loan and paid successfully in the last 12 months	250

Similarly, this type of detailed definition is available for 12–16 characteristics in a typical risk scorecard. This is one example of what a model looks like, but different software may produce different outputs, visuals, actual mathematical equations, etc.

There are two important concepts here: the maximum weight and the weight distribution across the possible values of the variable. Let's assume the total score for consumer risk is 1,000; the two variables we looked at had the maximum weight of 200 and 250, meaning they account for 450 out of 1,000, or 45% of the total score. The other variables will account for the remaining 55% of the score. So the first step of building a model is to identify the variables and their relative weights with respect to each other. The discriminatory power of a variable to distinguish good and bad records is used to identify the variables that will play a role in the model—that is, they will

Table 4.5 Risk Scorecard

Characteristics	% Weight	Maximum Score
Age	20%	200
Previous loans	25%	250
Income	15%	150
Residence	20%	200
Marital status	5%	50
Education	15%	150
Total	100%	1,000

become characteristics. As input variables from the training set are converted into characteristics by the algorithm to form the model, the relative weight or importance of the variable also emerges with respect to other characteristics in the model. It can be numeric, as used in the preceding example (regression-based models), or it can have probabilities representing their discriminatory power (decision trees).

Once the relative weight of the characteristic is established within the model, we have something that looks like Table 4.5. Scorecard is a numeric representation of a model.

These are characteristics, and each of them could in itself be a more complex derived or aggregated performance variable. In this example the loan decision is being made on this data from the applicant (and from credit bureaus) to assess the risk. This takes care of the maximum weight that each characteristic gets.

The second concept is that of the weight distribution across the various possible values of each characteristic. That problem is broken down into two pieces: the possible values and the weight of each value. For the age characteristic, we used 5 different possible values:

- 17 or below
- 18–24
- 25–45
- 46–62
- 63 and above

We could use 6 values or 10 values so why these 5? The rule of thumb for identifying the possible values should follow the same approach as discussed in the "Designing Performance Variables" section. The possible values are tied closely into the distribution of the overall population. Once the values have been identified, the weight distribution is based on what role each value has played in determining the 1 or 0 records—that is, the discriminatory power.

Data mining tools identify the weights themselves and provide as an output of the training activity as the model is built.

In regression-based models, this level of detail is abundantly clear—a statistician figures out the variables, their values, and their relative weights. Regression software packages also provide visibility into this level of detail but data mining predictive algorithms may not. It is alright to treat the data mining algorithm and the model it creates as a black-box. This level of detail is needed for tuning models, so if data mining software doesn't provide the detail, how do we tune the predictive models? We will deal with that later in the "Model Validation and Tuning" section.

Model and Characteristics in Descriptive Modeling

The descriptive models are different in nature from predictive models since they don't need to perform as accurately as the predictive models need to. Since predictions are for a potential future event and business wants to exploit that knowledge and take actions on the predictions, the reliability of the prediction matters a lot. A descriptive model, on the other hand, is describing the data in a form that allows for future action strategies, but it is not a precise event. Rather, it is a perspective into large quantities of data, so business can make sense of the data. It describes data in clusters or association rules so it doesn't need to be accurate, just approximate. Descriptive analytics has input variables, but their values and weights function differently. When a descriptive analytics model like clustering is complete or built, here is what you find:

- Number of clusters
- Cluster affinity (closeness of one cluster to another on the Euclidean plane)
- Cluster characteristics:
  - Cluster name or ID
  - Input variables and their values (or range of values) in each cluster
  - Probabilities and correlations of variables within each cluster

That makes the model and its characteristics in descriptive analytics much simpler to review, understand, and use. In predictive analytics, a future event is predicted and that has to be exploited favorably. The focus is the event and, therefore, the usage is tied to the event as well. In contrast, the descriptive model output is an explanation of the data using a structured form like clustering or social network analysis. Once the output is analyzed the question of exploiting this insight becomes wide open to interpretation and innovation. Therefore, the descriptive model's output is not expected to meet any fixed criteria. This distinction is further explained in the next section.

Model Validation and Tuning

Model validation is essential because of the black-box nature of machine learning algorithms. Once the algorithm has been acquired and data is run through it to build the model, it is important to know what kind of model was built. Is it reliable? Is it a good model? Can business decisions be made on that model? What happens if the model is wrong? The answers to these questions are more important, since this book is encouraging increased use of data mining to solve problems across all functions of a business. It is unlikely that functions of an organization, from HR to finance to procurement and operations, all have access to people who can mathematically review and validate the models. Depending on the data mining software or package, it may not be possible to fully uncover the workings of the algorithm, reconfigure the settings, and ensure good output, even if data mining experts are available on staff. Therefore, with the spirit of this book, we will present a simplistic method of validating a model to satisfactorily answer the questions just posed.

The model validation for predictive models is more critical than the descriptive models since predictive models directly influence a business transaction. The tuning is simply the process of getting the model to pass the validation test. The validation of analytics models is an approximation based on a benchmark of potential usefulness. If the model's use is not improving the results within the underlying problem statement, it needs tuning. For example, if a credit card default prediction model assigns "low risk" to customers who keep defaulting, then the model fails the validation. The same is true if the model keeps assigning "high risk" to customers who continue to be good customers. We will look at the predictive model validation first.

Predictive Model Validation

In a predictive model, since an event is being predicted, low False-Positives would imply the model is working. False-Positive is a term used to indicate the error rate of a model's predictions. A false-positive rate of 15% would imply that of the predictions made, the model was wrong 15% of the time. False-Positive counts both aspects of error in predicting an outcome i.e., when model predicts an event and it doesn't occur and when model doesn't predict an event and it does occur. However, if a probabilistic output of the predictive model is being considered on an outcome (estimation), then only high-probability outcomes of an event will qualify as an event and then their false positives will be evaluated. Let's use an example to explain this.

A shipping and logistics company moves boxes and containers from one location to another. They want to predict which shipments will get delayed. They take a sample set, let's say the last 12 months of shipping data, and they assign 1 to shipments that were delayed and 0 to shipments that made it on time. They ran 90% of that training data through a decision tree data mining algorithm and a model is built. When they tested the model with the 10% of training data that was set aside, they get a results matrix that looks like the following. Assume that 100,000 records were used in total so the model testing was carried out on 10,000 records while 90,000 were used for training.

10,000 Test Records	Shipments on Time	Shipments Delayed	Total
Actual outcome	8,610	1,390	10,000
Predicted outcome	7,840	2,160	10,000

- First inference: The model is 91% accurate for predicting shipments on time (7,840 out of 8,610).
- Second inference: The model's accuracy is 155% for predicting delayed shipments, or an error rate of 55% (2,160 instead of 1,390).

However, both inferences are misleading since we do not know the exact count of shipments that were actually delayed and have been predicted to be delayed or on time. So we build another matrix, called a confusion matrix, that shows where the model is actually wrong and where it is correct.

Confusion Matrix	Actual Shipments on Time	Actual Shipments Delayed	Total
Predicted on time	7,670[a]	170[b]	7,840
Predicted delayed	940[b]	1,220[a]	2,160
Totals	8,610	1,390	
[a]Shows where the model was accurate since the predicted outcome and the actual outcome were the same. [b]Shows where the model output was wrong.			

So:

- The total where the model is correct is 7,670 + 1,220 = 8,890 out of 10,000 (89%).
- The total where the model made a mistake is 940 + 170 = 1,110 out of 10,000 (11%).

The error rate is 11%, or the accuracy is 89%. The validation or evaluation of a predictive model is a science in itself. There are methods like the Kolmogorov–Smirnov Test (Kirkman, 1996), also known as the KS-Test, and Gini Coefficient (Gini, 1955), among various other methods to compare and evaluate predictive models. Explanation of these methods is out of scope for this book, as this is an introductory text designed for simplifying analytics adoption. The 89% accuracy of the model in our example above, is that good enough? Can the model be put into production? Let's look at some simple validation approaches.

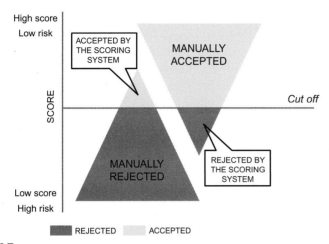

FIGURE 4.7
Predicted model decisions versus manual decisions.

Validation Approaches: Parallel Run

The simplest approach to evaluate the performance of a predictive model is to compare it to the non-analytics way of doing things. One approach is to randomly make some decisions using the predictive model and the remaining using the current business practices (which may be automated using some expert rules or are subject to manual review). Since we are still validating the model, the decisions that ran through the predictive model can still be reprocessed through the current manual process, but the results of the model should be saved for review in a few months to see if the prediction model results fare better than the manual decisions.

Let's take an example of a credit application system that has an embedded default prediction model:

- Input: The applicant's personal and financial information, credit report, and loan product parameters.
- Output: Credit risk score; a high score means low risk and a low score means high risk.
- Decision strategy: If the score is below the cutoff, reject the application.

Let's say 50% of the applications were processed through a manual process and 50% were processed through the model; the outputs were saved for validation purposes.

Management and business owners typically like this approach because it gives them better comfort and belief in analytics. Even though most of the inner workings are a black-box to them, the results shown in Figure 4.7 will be very transparent and convincing. The downside of this approach is that if the model was weak and didn't perform as expected and the results were not convincing, the analytics team will have to go back to the drawing board for another attempt at adding more variables or changing the existing ones. Usually there are three to five iterations before a model works satisfactorily, and it will take roughly 12–18 months to adopt a predictive model, as each model will need some quiet period to collect the results.

Figure 4.7 shows the applications that should've been rejected according to the model and the applications that should've been accepted. If after a few months of observation the manually accepted applications turn out to have defaults, then the model was right.

This approach has a side benefit, that of return-on-investment calculation on the analytics solution. If the money saved from prevention of defaults and the money that could've been made on rejected applications (that the model pointed out to be acceptable) added up for say three years is more than the cost of the analytics solution, then we know it is a good investment.

Validation Approaches: Retrospective Processing

In this approach we simply calculate the error rate or false-positive rate of the model from the 10% validation data and review its cost implications. In our shipping and logistics company example, the purpose of the model was better customer experience, retention and repeat business, and reduced fines or penalties when shipping SLAs were missed. If we add the cost of customer churn, lack of repeat business, and fines, that will be the total cost when a quantitative approach is not used. If we use a predictive model for identifying the potentially delayed shipments and prevent that cost from being incurred, we may create an overhead of 11% since the error rate was 11%. If that equation is favorable and makes good business sense, then we should adopt the model, otherwise we should figure out the threshold where the model will make sense and go back to tuning the model using additional or modified performance variables. On the other hand, if an existing practice of predicting the delays is in place using a subjective or nonquantitative method, look at its error rate, and if the predictive model has a better error rate, then the model is validated.

Model validation shouldn't be treated as an effort in scientific empirical proof. It should be reviewed in the context of the expected business benefit. As long as it delivers a business benefit and its false-positive rate doesn't negatively impact the business from its adoption, it is valid.

CHAMPION–CHALLENGER: A CULTURE OF CONSTANT INNOVATION

A champion model is the one in place that helps a business carry out analytical decisions. A challenger model is a fine-tuned version of the same model that is solving the same problem but using a different approach. The difference could be as simple as changing the design of an existing performance variable, such as instead of representing three months historical data for an event, it could represent six months' worth (TOT_PURCHASE_AMOUNT_LAST_6_MTHS). Or it could be as complicated as taping into an additional data source or changing the algorithm all together (e.g., going from a decision tree to a neural network).

As a guideline, for any model that is in production and actual business decisions are being made using the model, there should be a performance review quarterly or at most semi-annually. When off-the-shelf prepackaged models are purchased, the maintenance and tuning is typically priced separately. That cost should be factored into using analytics models. You either pay in building the internal capability to track and tune the performance of the models or you pay the vendor supplying the model. Skipping on that cost is dangerous, because it will not be known when the model is performing poorly and wrong decisions will be made.

Chapter 6 details how previous versions of the models have to be kept with the decisions performed using those models even when they are retired. How are various generations or versions of the model kept along with various challenger models that never proved better than the champion models? A careful audit trail is required for investigation and analysis when incorrect decisions based on models get highlighted by the business and a review is ordered.

The most important aspect of the champion–challenger constant improvement cycle is how do we know the challenger is better? Use the same technics described previously in this chapter in the "Model Validation and Tuning" section and see if the numbers from the challenger model are better than the champion model. If they are, you can move into promoting the challenger to the champion. However, that would require careful planning and coordination since various cutoffs of scores for decisions may need to be adjusted, and people and processes relying on the numbers from the models have to understand the change is coming.

When the challenger model takes ahold, start working on yet another innovative model using newer forms or newer sources of data and try it out to see the impact. Also track the performance of the model to know when it needs tuning. It is important to differentiate between tuning a model and replacing a model. Tuning is basically adjusting the input variables, while replacing means that a new approach to solving the problem is being considered. For example, on a propensity model where you predict the likelihood of a customer buying using a coupon or responding to a discount offer, you may have used several variables from the sales system and several from the call center and online store. Based on that, the model was built, and when it starts to deteriorate in its performance, you may tweak the variable definitions and add additional definitions. This is tweaking, because the underlying approach is that sales and customer interaction historical data is a good indicator of propensity. However, a challenger model would be where you introduce the human resource system variables. Now the underlying approach to solving the propensity problem is going through a fundamental change, where you are hypothesizing that some employee variables, indicating experience, education, salary, bonus, etc., have some impact on propensity and historical sales. This would be a challenger model.

This distinction is necessary because creating a new model has different cost implications versus tuning a model. There is additional overhead in data sourcing, integrating, identifying metrics, identifying candidate variables, etc. for building a new challenger model.

In Chapter 5, we will deal with how to handle the output of the model and convert it into an automated decision strategy.

Automated Decisions and Business Innovation

AUTOMATED DECISIONS

The purpose of analytics is to find patterns of data in an organization's data set. The reason for finding such patterns is to use them within the normal business activity to help maximize value from business operations. This chapter covers how to go about embedding this approach into the DNA of an organizational and management culture. It will break down the foggy marketing mantra "use analytics to maximize business value" and build a step-by-step approach to demystify and make this goal achievable.

There is a school of thought within analytics professionals to use analytics to empower superanalysts dubbed data scientists (Davenport, 2012), give them all the power of the analytics models, and let them figure out how to use that model output for maximum business value. The approach emphasized throughout this book is the democratization of analytics, not limited to a handful of problems where data scientists may be available. Data scientists are required to have skills in data and analytics as well as deep insights into the business, which is a very rare combination to find. Therefore, business decisions made with the help of a data scientist have to be limited to a handful of functional areas where such skill is available and affordable. The automated decision approach democratizes the use of analytics in business decisions by converting the subjectivity into an objective, well-defined, and transparent set of rules under a decision strategy. How to devise these rules and ensure they continue to deliver value from the analytics models is covered later in this chapter in a section on strategy validation and tuning. Whether this decision strategy rules–based approach is superior to the data scientist approach will be addressed in the last section of this chapter on business process innovation.

DECISION STRATEGY

The takeaway from this entire book is actually buried into this chapter—dealing with and building effective decision strategies. Large and sophisticated data storage and data processing infrastructure, state-of-the-art algorithms or software packages for analytics, and Big Data toolsets—none of that matters if the decision strategy component is not properly designed, managed, and executed. By definition, analytics has a degree of gray area unlike a report run on historical data that exactly represents what has happened. This gray area requires a certain degree of flexibility to accommodate varying situations, scenarios, and options. Not all of these scenarios can easily be accommodated in a model since then each scenario will have to have its own model and not all scenarios will have enough training data available to build a meaningful model. Besides, if the scenarios are not mutually exclusive then the overlapping models with conflicting outputs become a very complex equation to solve. The scenarios are also dictated by market-changing conditions, competitor's actions, and even environmental disasters. It is not possible to build models to cater to all that. Therefore, a model is built on historical data assuming "business as usual," and then the dynamic scenarios are handled with decision strategies as to what should be done in certain situations given that the business knows what the model says.

Nassim Nicholas Taleb in his book *Fooled By Randomness* (2008) and later in a revised work in *The Black Swan* (2010) talks about the weakness of models when a surprising event (i.e., a black swan) occurs and renders the models useless. Surprise events will occur that have not been addressed by the historical data used to build a model, but the key is not to abandon models or be overly reliant on them. The trick is to find the balance in decision strategies to be able to adjust in near–real time during unexpected times, limiting the damage from automated decisions. It is not possible to retrain the model with the surprise event factored in because the recency factor and limited set of historical data for the surprise event will probably get skewed in the training, and by the time model is rebuilt, tested, and validated, the damage is probably done. A decision strategy, on the other hand, can be easily turned more conservative or aggressive depending on the circumstances or business objectives and allows for manual intervention instead of resorting to default decisions.

Earlier in the book we talked about the mortgage crisis of 2008 and the fact that risk models had correctly dubbed these risky customers as subprime, highlighting their high probability of defaulting on their mortgage loans. The analytics model was correct but the decision strategy was aggressive and led to a massive crisis. However, the mortgage-backed securities risk models that called them AAA were actually the problem of the models that had not factored in various performance variables that in hindsight we know should have been factored in. In that scenario, the strategies probably worked fine assuming the risk indicators were accurate from the model. There is no magic wand to build a system and allow it to continue to create value while we sit back, relax, and enjoy. Both the models and the decision strategies have to be carefully designed, executed, monitored, and tuned almost as a managerial life style of the Big Data era. However, understanding models, their inner workings, and the mathematics of algorithms is far more difficult for a midlevel manager or an executive versus managing the decision strategy, which is based on business transactions and operational flow. The input of the analytics model into a business decision has to be managed through the decision strategy, as that is where the real value of analytics comes to fruition. Understanding and managing a decision strategy comes naturally to a business operations manager as we shall see in the rest of this chapter.

Business Rules in Business Operations

There are business rules in every business operation responsible for carrying out the day-to-day transactions across sales, service, procurement, hiring, etc. These rules make up the decision criteria used by the business to identify and resolve situations maximizing value in favor of the business. All of these business rules have numerous decision variables (DVs) that need to be defined to complete a transaction. The DVs in the following examples are in double quotes:

- If you want to offer a discount to a good customer, how to define "Good."
- If you don't want to lend to a customer with poor credit, how to define "Poor."
- If you want to reroute a package using a premium service otherwise it may get delayed, how to define "May" get delayed.
- If you want to stop a money transfer as a suspicious money laundering transaction, how to define "Suspicious."
- If you want to hire the most suitable candidate for a position, how to define "Suitable."

All business operations deal with these types of questions. The answers to these questions are extremely important as a policy enforced on the employees in the trenches carrying out dozens if not hundreds of such transactions in a day. These are called business rules. What to do in a certain situation? How to make a decision? How to maximize the business value of each business decision? These and thousands more are questions that business rules answer. Business rules are well defined explaining the scenario and the response. There are two primary categories of business rules: expert and quantitative.

Expert Business Rules

Expert business rules, as the name suggests, are devised by people who have decades of experience in a certain line of business. Their specialization in a particular area makes them experienced in all kinds of situations and scenarios, and they understand the impacts of the business decisions on the bottom line very well. They become the policy designers, mentors, and go-to people for other employees to get advice on how to handle a certain business transaction.

Most straightforward and business-as-usual rules (80% of normal business) are either baked into the operational systems or the employees are well trained to identify and address them. It is usually the 20% that really need an expert input. The experts also have some degree of leeway in their decision-making authority and are able to waive fees, offer discounts, approve cases, etc., as long as it is in the best interest of the business overall in their opinion.

They usually have their own standards to define DVs, Good, Poor, May, Suspicious, and Suitable. These definitions will vary from one expert to another driven from their education, training, experience, and general approach to their profession. When these experts leave an organization, they leave with a lot of institutional knowledge on decision making in unique scenarios.

Quantitative Business Rules

Quantitative business rules are based on hard numbers and absolute values for decision making. For example, a retailer may define a Good customer as someone who has been a customer for over one year and has purchased at least $300 worth of products or services. A bank may define a money transfer that is over $7,500 to be suspicious for money laundering if the customer has maintained an average monthly balance in the account of less than $200. The reason why these types of rules are quantitative is because these DVs have been arrived at after extensive analysis of historical data and these values were found to be the best cutoffs for these business decisions.

Decision Automation and Business Rules

When business operations get insights from historical data through the analytics models, the state of these business rules changes because they now need to be reviewed in light of additional knowledge. On the other hand, because of these new insights, new rules will be needed to put the insights to work. These new rules can be subjective or expert-driven in some unique situations where enough historical data is not available, such as using Facebook activity to assess the suitability of a candidate for a job or Twitter activity as a basis for identifying premium customers. There are other areas of machine and sensor data that have been brought into the analytical space fairly recently and therefore not enough quantitative evidence is available. Some examples of this are additional sensors in cars that record and report on driving habits or smart meters for electricity and gas replacing the decades' old analog meters.

The business rules that make up the decision strategies have to follow a structured and documented process regardless if they are expert or quantitative. Quantitative rules should be preferred but the rule creation or design process should carefully look at the analytics problem statement, the model's output, and the context of the decision being made to identify the DVs and the appropriate cutoffs.

Joint Business and Analytics Sessions for Decision Strategies

The design of decision strategies requires input from business for two main reasons:

1. They understand their business processes and rules very well.
2. They need to alter their business activities driven from analytics so their buy-in is needed.

It is very unlikely that the team that has built the analytics model would actually understand how to put it to work for business improvement. Just the fact that you can reliably predict which customers are likely to defect, which loans are going to default, or which products always sell together doesn't mean you also know what to do about this information. Therefore, joint sessions are needed where the results of model validation should be shared with business. Explain to the business units what patterns in historical data do for future trends and then pose the question as to how this would get utilized.

Drawing a treelike structure with some arbitrary choices of DVs and their cut-offs will get analytically savvy business people excited and they will soon be building complex decision rules and strategies. They should be able to build these strategy trees in any flowchart-type drawing tool like Visio or PowerPoint.

The following examples will explain decision strategy design in more detail.

Examples of Decision Strategy

We will use two very famous and well-documented examples of decision strategy to help describe the rules that make up the decision strategies, the DVs, and their values used for decisions, and how to validate, manage, and tune a decision strategy used to make automated business decisions on analytics models.

Retail Bank

A midsize retail bank has a car loan lending product for consumers. So consumers apply for a car loan with the bank and the bank, after due diligence, decides to either accept or reject the loan application. If it accepts the car loan application, the bank funds the customer for the car loan and then manages the loan servicing over the term period to receive payments and earn interest income. Analytics is heavily used in this business to determine the risk of default on that loan. So the bank needs to know the probability that a certain customer requesting a loan will not be able to pay. Here is what a bank needs to do:

1. The bank takes the loan data from its historical data set and identifies the loans that were properly repaid and the ones that defaulted.
2. The bank also gets predictive analytics software to build a predictive loan default model.

3. The algorithm takes the data set as an input (uses 90% of the data to build the model), identifies the patterns in data that differentiates a good loan versus a loan gone bad, and trains the model.
4. The model is tested using the 10% of the data that was set aside for this purpose.
5. The model is tuned as necessary (addition, removal, or redesign of the performance variables used in the model) until it is ready for production.

These steps are common for every analytics modeling problem and every type of analytics technique, its algorithm, and its model. Now that the model is ready, decision strategy comes into play. Let's say the model is implemented in standalone software and loan applications come in through a web-based system. As a loan application comes in, it is passed to the software housing the analytics model and the model returns a probability of default back to the application processing system. This is done instantaneously so the customer may actually be sitting in front of a web page awaiting a response. What should be the response that can be sent back to the customer in real time? Do we show the probability coming back from the analytics software? That would not mean anything to the customer, so somewhere this probability has to be translated into an appropriate "Congratulations! Your loan has been accepted" or "Sorry! Your loan application has been denied" message. This is the job of the decision strategy. The risk and credit officers at the bank have built a strategy around the probability that is received from the predictive model.

The strategy includes several decision variables and their thresholds for decision making. In this example typical decision variables include:

1. Probability of default, between 0 and 1
2. Loan amount requested
3. Down payment

There is an additional set of policy variables where certain values mean outright rejection, such as convicted criminals, falsified records, identity theft or fraud indicators, age below the policy requirement, etc. The decision strategy for the loan approval will be as follows:

```
IF any of the Policy Variables are TRUE
THEN Reject
ELSE
    IF the Probability of Default > 0.62 (62%)
    THEN Reject
    ELSE
        IF Probability of Default is between 0.28 and 0.62
        THEN IF Loan Amount Requested <= 10,000
            THEN Approve with 25% down payment
            ELSE (i.e. loan amount requested is > 10,000) Request
            for a Co-Signer
        ELSE Approve the loan (i.e. probability of default is < 28%)
```

It should be obvious from this example that analytics models and the predictions alone are not enough to get value out of analytics. The actual business decisions require additional consideration of various unique situations and scenarios. In a user-centric decision environment, this would be left to an employee to weigh in on gray areas and use good judgment. However, that introduces subjectivity and differences in approaches leads to inconsistent decisions and results when analyzed over time.

It is important to note here that the performance of the analytics model is measured on the results taken within the decision strategy. If the decision strategy has inconsistent and subjective decisions, their outcome can incorrectly evaluate the performance of the model. Some employee may make good judgment calls on what decisions to make and hide the weaker predictions from the model and vice versa. Therefore, it is important to use automated decision strategies and implement the subjective rules into absolute values so the variation in decision strategies is eliminated when evaluating models.

The performance of a predictive model is measured using the results of decision strategies, therefore it is important to isolate their influence on models' outputs.

Decision Variables and Cutoffs

This example uses the following DVs:

1. Probability of default (with values 0.28 and 0.62)
2. Loan amount requested (with a value of 10,000)

the probability of default DV comes from the risk predictive model, while the loan amount requested DV is part of the business transaction where analytics is being applied for a decision. The values for the decision variables have been analytically derived. High default risk means "Bad" customer so the definition of Bad is basically greater than 62% probability of default. This definition has been arrived at looking at historical data and reviewing past defaults. The same goes for the other values within this decision strategy. Figure 5.1 depicts a visual representation of the decision strategy.

Insurance Claims

The insurance claim example we use here has multiple possible decisions unlike the loan strategy that had only two possible decisions, approve or reject the loan. The insurance industry is perhaps the oldest user of analytics with actuaries trained and specializing in the discipline of assessing future outcomes of various events. Different types of insurance, such as property/casualty, life, and healthcare, all have a similar problem of calculating the probability and size of a claim on an insurance policy. This calculation

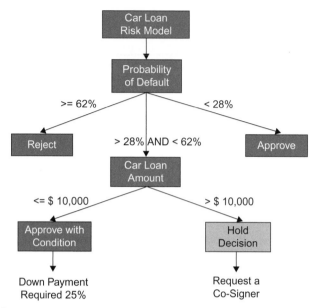

FIGURE 5.1
Representation of the decision strategy.

dictates how much premium to charge a customer. A predictive model can be
built that can predict the probability of a claim on a policy, but the size of the
claim will be assessed using the decision strategy, which will lead to a decision on how much premium to charge. Figure 5.2 shows a decision strategy
that uses a predictive model to provide the probability of making a claim on
a policy, and then a set of rules to determine the premium.

Decision Strategy in Descriptive Models

In descriptive models, decision strategies are still needed to address the gray
area introduced by descriptive models. In case of outlier detection using
clustering, the idea is to supply a large data set to a clustering algorithm
and it will plot the data points and look for points that are close to form a
cluster. Once the clusters are formed and their definitions have been identified, the descriptive model is ready. When a new observation comes in, it is
determined which cluster it is part of *or* if it is close to a particular cluster. A
decision strategy or the clustering software can compute that looking at the
ranges of variables within each cluster and the values on the new observation. Once it is determined whether the new observation is in the cluster,
the expected future behavior of the new observation will be the same as the
rest of the population of the cluster, and therefore a decision can be made
about decision on the new observation. On the other hand, if it lies outside
a particular cluster, then it must be determined how to apply the insight of

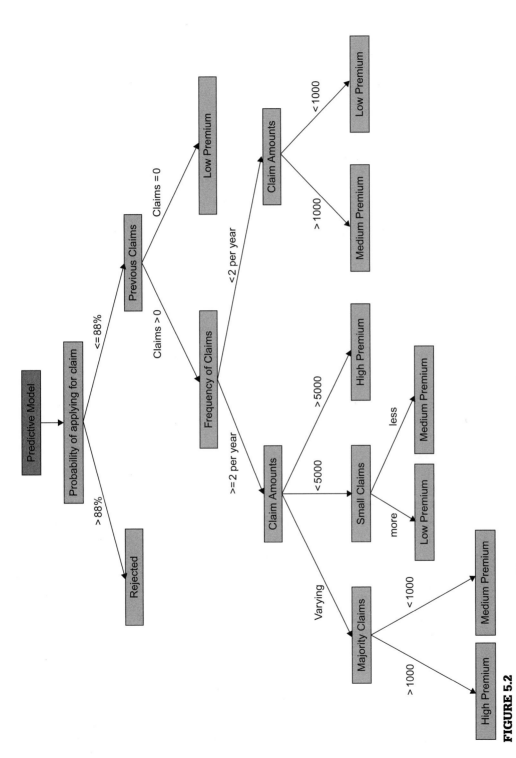

FIGURE 5.2

Decision strategy using a predictive model.

the cluster it is closest to. The Euclidean distances of "in the cluster," "outside the cluster," and "how far outside the cluster" are all gray areas that need a decision strategy to determine how to treat them.

Similarly, in a social network analysis, if the relationship between two customers is too strong, a certain type of decision is warranted, however, if the relationship is somewhat strong, then a different strategy is required. This again introduces a degree of gray area that requires business rules–based decision strategies. The decision strategies may be a little more volatile initially and they may go through several iterations and adjustments before getting finalized. Apart from that gray area in a model's output, everything that has been covered in predictive decision strategies applies to descriptive analytics–based decision strategies, as the actions on the analytics' models are still a set of DVs, their thresholds, and business decisions.

DECISION AUTOMATION AND INTELLIGENT SYSTEMS

Now that we have seen the process of how operations run, generate data, and the data is stored and analyzed, analytics models come into play. The use of analytics models is through decision strategies. Now let's look at how these strategies go out of the analytics team and into the real world so actual business operations benefit from analytics—the culmination of value from data through democratization of analytics.

Learning versus Applying

The purpose of data warehouse and analytics systems is to analyze data, learn from that analysis, and use that knowledge and insight to optimize the business operations. The optimization of business operations really means changing the business processes so they yield greater value for the organization. The business processes are actually automated through operational systems, so these changes require operational systems changes. However, since we established in Chapter 1 that analytics has to do with a future event, the business process changes are limited to well-defined events and the response to those events. Just to ensure that there is no confusion, modifying or improving a business process can also mean improving the information flow in a process, the integration of various disconnected processes, or eliminating redundant or duplicate steps from the flow. That is, the field of business process management has nothing to do with analytics models and decision strategies. The business process optimization, redesign, or reengineering within the purview of analytics is limited to the automated decisions driven from business rules that take input from analytics models. These business rules are embedded in the operational system, and therefore the application of analytics input has to be implemented within the operational system or as an add-on component or extension of the operational system.

Decision automation has two dimensions: the learning and the application of that learning on business operations. The learning is where the data warehouse, data analysis, and analytics models come into play, while applying that learning on actual business activity is where decision strategies come into play, and they have to be embedded or tightly integrated into the operational system. There is a school of thought known as active data warehousing that suggests that this decision making should be done in the data warehouse since all relevant data is available to make a determination comprehensively, from all perspectives. This requires the data warehouse to be connected in real time to the operational system for receiving the event and responding with a decision. This just increases the complexity of integration, as the rules or knowledge used to make the decision must have been done beforehand, so why integrate with the data warehouse in real time? The active data warehousing approach works well in campaign management systems where the line between operational data and analytical data is blurred. It is not a recommended approach when additional layers of analytics are involved. If the decisions are totally based on some triggers or thresholds that the data warehouse is tracking in real time, then active data warehousing may work. The strategy integration is a superior and simpler approach and decouples the data warehouse from live events and decisions. The results still go in the data warehouse within the analytics datamart, but there is tighter monitoring control and a simpler interface for strategy modification and testing.

Figure 5.3 represents this learning versus applying. If we break this diagram in the left side (going top to bottom), that would be the learning dimension, and the right side of the diagram would be the applying dimension. From a system architecture perspective, the nature of the two is quite different, and

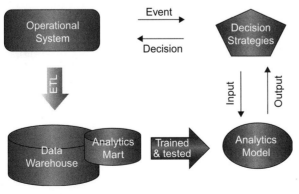

FIGURE 5.3
Decision strategy technical architecture.

therefore a modular approach allows for greater flexibility in choice of tool-set, choice of monitoring and control, and the operational SLAs to support the business.

Strategy Integration Methods

Earlier in this chapter we used a simple decision strategy example for a consumer car loan. That example was presented in its algorithmic form (a combination of nested IF_THEN_ELSE), as well as in visual form (see Figure 5.1). Looking at that simple business rule it should be obvious that adding that extended logic in the operational system workflow is not a technical challenge at all. The analytics tool will always stay outside as a black-box or a standalone component. However, if there are several scenarios that need their own strategies or there is a complex situation where multiple models are also involved, this type of extended coding within the operational system may get too complicated to maintain.

On the other hand, a pure-play fancy strategy management tool may be too expensive. Therefore, one method is to embed the strategy rules in the operational system software. Another is to buy a strategy tool. The downside of embedding the code is that monitoring and auditing, as well as modifications, testing, and what-if scenarios, will be extremely difficult to carry out. The strategy tools do solve this problem as they have visual interfaces for strategy design and internal version controls, but there is the integration with the analytics tool, data warehouse, and operational system that has its own costs and implementation overheads. Here is an innovative approach as an alternate method to the two methods described earlier.

ETL to the Rescue

ETL (extract, transform, and load) is a small industry comprising specialized software, comprehensive data management methodology, and human expertise existing within the data warehouse industry. It came into being with the data warehouse since data needed to be pulled out of the operational systems and loaded into the data warehouse systems. We refer to ETL as a noun encompassing all aspects of "data in motion," including single record or large data sets, messaging or files, real-time or batch data. ETL then becomes the glue that holds the entire analytics solution together.

All ETL tools now have GUI-based development environments and provide all the capabilities of a modern software development tool. If we look closely to the two treelike strategies in Figures 5.1 and 5.2, they look very similar to how a data processing or dataflow program looks in ETL tools. Therefore, an ETL tool can be used to design and implement strategies. ETL has its own processing server and has integration interfaces to all source systems; there is plenty of expertise available within all data warehouse teams. So the recommended integration is through an ETL layer that receives a real-time event from the operational system, it prepares the input record around the event, and invokes the analytics model. The model returns an output that ETL will take into the strategy, run the data through the strategy, and reach a decision. It will then deposit that decision back into the operational system. This is a very simple integration for ETL developers who have been building the information value chain through the Information Continuum. Remember, if the prerequisite layers of the hierarchy are not in place, jumping into the analytics levels of the Information Continuum will only yield short-term and sporadic success. For a sustained benefit from analytics, the Information Continuum has to be followed through each level, and that will automatically bring the ETL maturity needed to implement decision strategies and integrate with the operational systems.

STRATEGY EVALUATION

The evaluation of decision strategies serves two purposes. One is to validate the automated decision making when it is being put in place for the first time and the second is when an alternate and competing approach is being tested. There are two primary methods for decision strategy evaluation: retrospective processing and reprocessing.

Retrospective Processing

The retrospective processing method is identical to the approach used in model validation and covered in Chapter 4, and therefore uses the same name. The historical data, such as customers or transactions where the strategy should've been applied, already have a known outcome. If the business value from those prestrategy decisions on such transactions is known, provided the same model was used, then we can do a fair comparison by running those records through the strategy and, based on the model's output, reach a decision. We can then compare the business value of these decisions.

Reprocessing

A more prudent approach is to run the same event or transaction through the decision process that is currently in force. It could be a champion strategy in place, or it could be some subjective decision rules driving the decisions on those events. At the same time, randomly select 10% of the transactions and run through the new decision strategy and save the results. After some time, when the true benefit of the decision is evident and measurable, compare the results of the two approaches. More robust decision strategy systems, such as the ones available from SAS or SPSS for general-purpose decision strategy management or specialized ones from Fair Isaac (FICO) or Experian designed for consumer lending business, have the ability to create a simulation environment to even test what-if scenarios with manipulation of strategies. For example, if the total cost of the decision strategy used to keep customers from defecting through discount offers is $3 million and the value of their retained business is $4 million, then a simulated environment can show what that number would be if the criteria or the value of the discounts was adjusted. If the cost of the simulated decision strategy comes out to be $2.7 million while the retained value is $3.9 million, then it is possible this discount strategy should not be used. This means that the criteria for offering discounts was changed in the simulated strategy and not all the customers got the discount as before, but yet again the benefit was almost the same.

CHAMPION–CHALLENGER STRATEGIES

Business should be monitoring execution and effectiveness of strategies on a regular basis. The analytics datamart that stores the events or transactions that came in and were acted upon, directed by the decisions coming from the strategy, are recorded with an intimate level of detail, including the input, output, and the path it took down the strategy decision tree. Reports running against this data should allow users at all levels in and around the business process to look for new ways to use the analytical insight and constantly improve the processes. It may not be a bad idea to mandate that all concerned should come up with challenger strategies (may be incentivize as an internal competition) and at least one challenger strategy should be selected to compare against the champion strategy currently in force. The strategy evaluation methods can be used to evaluate the performance of the challenger strategy and, if it yields superior results, it should get utilized. The employee responsible for that strategy should be recognized and compensated accordingly.

Business Process Innovation

The constant and sustainable cycle of innovative strategies will improve business processes all across the organization with simultaneous benefits of operational excellence, product innovation, and customer intimacy. The defining impact of democratization of analytics on a business is that innovation is no longer top-down driven from a handful of brilliant executives. Rather, all employees across all aspects of the business are able to improve the business operation they are responsible for. Procurement, supply chain, HR, sales, marketing, and customer service all get their own models, their own strategies, their own process of champion–challenger models and strategies, and an innovation and reward culture. That is how we enter, survive, and excel in this era of Big Data.

Earlier in this chapter we briefly reviewed the widely touted analytics approach of using a data scientist and explained some of the challenges of that approach. The last few sections in this chapter have provided a detailed explanation of an alternate approach—democratization of analytics. It is up to the individual organizations to understand the intellectual capital within their technical and business teams and see if a data scientist approach is feasible for them. For some handful of specialized areas it will always be a superior approach, but the strength and depth of democratization of analytics cannot be ignored. An organization that can adopt the culture of data and analytics across all their business processes can certainly stay a step ahead of competition and may operate with more efficiency, produce improved products, and serve their customers better by relying on their data and the collective genius of their employees.

Exercises

Following are a couple of problems that anyone should be able to attempt even without formal domain knowledge.

5.1 Student drop-out prevention. Build a decision strategy that uses a predictive model that computes the likelihood of a student dropping out. The strategy should be designed to prevent the student from dropping out.

5.2 Churn prevention. Build a decision strategy that uses a predictive model that computes the likelihood of a customer defecting (leaving a monthly subscription service like cable TV, cell phone, etc.). The strategy should be designed to keep the customer from defecting.

CHAPTER 6

Governance: Monitoring and Tuning of Analytics Solutions

Audit and control is an essential component of analytics solutions and is required to be in place before business decisions are carried out using analytics. The purpose of analytics is to stay ahead of actual events and have some future perspective and predetermination of business transactions. Since analytics enables proactive actions on events as soon as they happen, it is important to have a tight monitoring on those decisions to prevent damage from mistakes, abuse, and even incompetence. The tools and approaches for audit and control of analytics-driven automated decisions are not very different from the ones needed by owners of the models and strategies who track their performance. The only difference is that the audit team has to be outside of the typical IT and business groups responsible for tuning and modifications of models and strategies.

ANALYTICS AND AUTOMATED DECISIONS

Here is how analytics enables automated decisions as we have seen in the previous chapter:

- Regular business operation carried out by operational systems generates data.
- Data is accumulated over time to build history.
- Historical data is used to develop insights into what has happened and how business has been performing.
- Analytics models are built from historical data to influence future business decisions.
- Decision strategies are built on analytics and integrated with operational systems for automated decisions, with minimal to no manual intervention.

The benefits of this entire approach are well articulated, well understood, and widely adopted as described in previous chapters. Some of them are:

- Superior efficiency and speed compared with the manual process.
- No time and geography constraints—systems are up and running at all times and looking into all aspects of the business across all time zones. Humans have serious limitations with time and geography.

- Real-time and event based—once a system detects an opportunity, it follows up and executes the decision accordingly.
- There are no hidden agendas, ideological beliefs, and biases when a system automatically makes decisions.

However, there are some downsides of this approach that are not as widely understood and addressed:

- The analytics models and their accuracy are dependent on the input variables. If the input variables become irrelevant because of changing business conditions, the models become stale and their output may not be as reliable, but the automated decisions wouldn't necessarily know that (e.g., AAA-rated mortgage securities during 2008 economic crisis; Financial Crisis Inquiry Commission, 2011).
- The conditions or scenarios as laid out in the decision strategy have to occur exactly as designed for the automated decision to work. If there is a slight bit of difference in the real-time event compared with how it has been lined up in the strategy, the decision will not take place as intended. The default conditions and ensuing decisions can therefore be unpredictable, as not all possible real-world situations can be anticipated and addressed.

The Risk of Automated Decisions

The downside of analytics-driven automated decisions is the risk of incorrect decisions that are carried out automatically without detection until it is too late. There is plenty of evidence that automated decisions in trading of equities have suffered from this symptom where the systems kept making the trading decisions, creating a snowball effect and compounding the negative effect (losses). These trading decisions have analytics models and decision strategies behind them covering various scenarios, and they make automated decisions without human supervision. There are automated decisions that if incorrect can result in losses.

Another risk is inaction leading to lost opportunity, where the scenario was slightly different than the one coded in the strategy and therefore no decision was made.

Another risk is wasted resources, where a remedial action was taken, such as refinancing a loan or rerouting a shipment, when it was not required in the first place (this is mostly because of false positives). Does the risk of incorrect decisions outweigh the benefit? Should we stop using decisions based on analytics models and automated decision strategies?

Monitoring Layer

Instead of adopting an either-or approach (use automation or not in decision making), we will try to build a monitoring layer on top of the analytics-driven strategies to mitigate the risk of incorrect decisions. The rest of the chapter is dedicated to details behind building and using this monitoring layer.

AUDIT AND CONTROL FRAMEWORK

The audit and control layer is designed and built the same way as the reporting and analytical layers within the Information Continuum. Operational systems carry out the business operation and generate data. The data is used to measure and keep an eye on the business. Now that analytics-driven automated decisions are being used, it is yet another form of business operations carried out by decision strategies integrating into the operational system. Therefore, a new reporting and measurement mechanism is needed. Automated decisions have two parts: the part relevant to the analytics and the strategy, which includes information such as:

- Analytics model name/ID, type, and version used in the decision
- Decision strategy name/ID and version
- Input record values (the input variables required to run the model)
- Output from the model
- Decision path within the strategy
- Input date/time, source system for the input record, ID used, output date/time for the decision, etc.

This information is actually the metadata about the automated decision. The other part is the real data of the transaction, such as approving a case, sending a coupon, buying a commodity, etc., and that data still lives in the operational system and is recorded in the same manner as before—it is carried into the data warehouse and reported against. That is, the actual data is recorded in the operational system and the traditional data warehouse system. The analytics decision metadata is recorded in the analytics datamart. The audit and control information is within this metadata in the analytics datamart. We will see what information is relevant and how to use it.

Organization and Process

The audit and control organization has to cut across various analytics projects. The idea of adopting analytics in the manner outlined throughout this book is to make it available to all business functions across an organization. That requires audit and control monitoring to be in place within each business function that is carrying out automated decisions using analytics models. Instead of trying to create that role within each business function, a centralized audit team ideally within the internal audit department should be built. There is nothing special about the audit of automated decisions, and any trained auditor should be able to pick up the new role. Most of the audit activity has to be automated anyway.

The process that the auditor has to follow is based on standard reports that show summaries and trends of decisions being made. The data is moved into the audit datamart from the analytics datamart and then the reports run and provide visibility to the auditor about what is going on. This is a reactive mode and will be needed in the early stages of analytics adoption. Once the automated decisions have been in place for six months or so, there will be some patterns and trends available that can allow setting up thresholds, alerts, and triggers. The auditor will be able to set these up and tune them. Some will be set up just for monitoring purposes and some will act as a circuit breaker to override the automated decisions being made.

Audit Datamart

The audit datamart design will follow the design principles as widely understood and used within the business intelligence and data warehousing space. The audit datamart is a classic datamart designed for auditing users who have an interest in specific audit data to be extracted from the analytics datamart and the data warehouse and loaded into a separate auditing datamart. Figure 6.1 shows an illustration of the audit datamart. Only analytics-driven automated decision entities are presented, and it is expected that each organization will introduce its own flavor of audit to this data structure.

The decision summary fact table records the decisions being made by the decision strategy integrating in the source system. Its grain can be at each decision level, or if bulk decisions are being made, then it will have aggregate data. The decision code is the decision that the strategy eventually takes. For most predictive modeling solutions and strategies, there will be typically three decision codes. For example, in case of a loan application, the three decisions can be accept, reject, and review. The decision factors in the predictive model's output that assigns a probability of default, and then the strategy looks at the overall situation of the applicant, the loan offer, or the product specifics, liquidity, and lender's overall strategy, etc. to make a decision, but the overall possibilities of a lending decision can have just three codes. The threshold and alert tables are not the typical datamart tables. The threshold table is set up by the auditing team with respect to automated decisions. They can put thresholds on certain decision code or allow only certain counts or values of decisions to go through. Once a threshold has been designed and inserted into the threshold table, any breaches will get recorded in the alerts table. These are broad ideas to consider when designing an audit datamart for analytics.

Like all datamarts, this process will use all data warehousing best practices, such as the use of an ETL tool to load data in the audit datamart, regularly

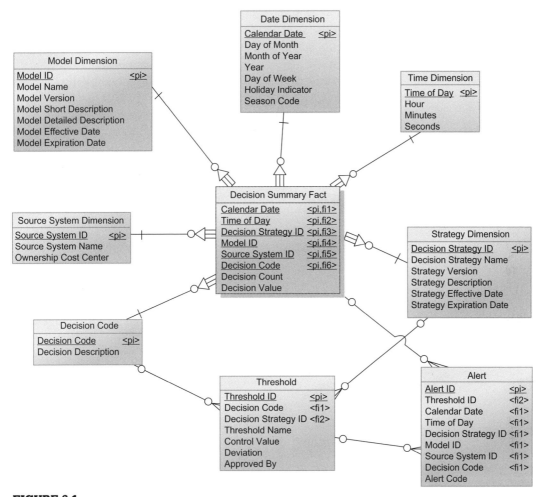

FIGURE 6.1

An audit datamart.

scheduled data feeds, data-quality controls, etc. In addition to the data management for the datamart, the front-end or the presentation layer of the audit datamart has to follow suit. The same reporting tool used by the rest of the data warehouse for reporting should be used for the audit reporting.

Unique Features of Audit Datamart

The audit datamart will have some uniqueness in its implementation compared with the other datamarts within the data warehouse:

1. The loading, backup, and archiving schedule will have to be shared and signed-off by a member of the auditing team.

2. Any data discrepancy or data-quality issue will also go through a more rigorous investigation.
3. Any changes to the datamart will have to be signed-off by the auditors.
4. Separate logs will be maintained for users accessing the datamart and running queries and reports. The logs will only be accessible by the auditors, and they will decide when to discard or archive the logs.
5. The presentation layer accessing the audit datamart will have tight security and access controls. Most BI tools have robust security and access controls and they should be used to limit the access to the audit datamart.
6. The threshold table will have a separate load interface controlled by auditors to change various thresholds or remove or add new ones. A small application with appropriate security controls may have to be built.
7. The alerts table can be set up to run nightly to look at the data for the day, check against thresholds, and generate an alert when a threshold is breached.

Control Definition

Designing the thresholds is a complicated task. It involves clear understanding of the business process, the problem definition for which the model has been built and used, the various scenarios and decision parameters used in decision strategies, as well as the risks of incorrect decisions. It is very unlikely that an organization has this knowledge readily available within a team or department. The auditors by design are required to maintain this level of understanding for critical functions within the organization. However, we are proposing democratization of analytics-based automated decisions and not limiting to a handful of functions around financial transactions. The purpose of highlighting this issue is for the management and decision makers, essentially the champions of business analytics within the organization, to understand the challenge of auditing automated decisions. There are three methods of designing the control that are presented next, and it is up to the individual organization which one provides the best alignment with their internal audit organization's culture.

The underlying idea in all three methods is that the process of auditing should be automated as the decisions are automated. If the decision monitoring is automated, then controls and thresholds are needed to put in circuit breakers that trip the automated decision process.

Audits are needed for analytics-driven automated decisions to introduce the circuit breakers tripping the automation and protect against incorrect automated decisions.

Best-Practice Controls

The best-practice controls are controls that can be copied from published case studies or other guidelines shared in courses, trainings, and marketing material. These are typically available for a specific industry and use. If we look at the automated decision controls and circuit breakers that have been put in place at the New York Stock Exchange, they were in response to an event in May 2010 that came to be known as "Flash Crash." Learning from that mistake of allowing automated decision systems to keep trading in stocks, most trading exchanges have put in monitoring controls with appropriate circuit breakers that get tripped, and the source of suspicious trading orders is identified and excluded.

It will be awhile before such case studies will be widely available for other industries and problem domains, and therefore the best practices that can be followed will be driven from parameters to look for and building an internal control. For example, an analytics-based automated decision system put in place that looks at a customer profile or a customer activity and offers a discount on another product. There are no guidelines as to how many such discounts should be offered. What if too many customers show up on the website and receive discounts? However, the best practices are available to aggregate the discounts until a budgetary threshold is reached. Similarly, for a credit card online-approval automated decision system, there is no best practice on how many applications should be approved, rather the liquidity and risk-adjusted capital ratios that are well established within a lending operation act as a threshold. Existing controls and thresholds should be used initially and then improved over time adjusting to the unique nature of automated decisions.

Expert Controls

The expert controls, as the name suggests, are controls designed by experts. Designing expert controls is very similar to designing expert rules in decision automation strategies. Domain knowledge and years of experience allow an expert to build some subjective controls that they use in their day-to-day activities to manage operations. Experts should be interviewed extensively to understand what triggers their curiosity in a certain situation and how they react looking at the changing business situations. The reports that they review from the data warehouse should provide plenty of clues as to what pieces of information are critical to their decision making and should act as controls.

There are no specific guidelines on expert controls. If domain experts are available who are data-savvy and have been using good judgment in utilizing their subjective controls in the past, use that knowledge and convert it into a formal implementation of controls through a system of triggers and alerts within the audit datamart.

Analytical Controls

Analytical controls deal with building a "business as usual" scenario and flag the decisions that do not fit the profile. The establishment of this scenario requires historical data analysis very similar to how decision variables and cutoffs are used when quantitative rules are being devised within decision strategies. Some examples are:

- Typical volumes of automated decisions expected and actually carried have to be part of the controls, so if typical volumes exceed by more than 20%, someone is alerted. If volumes exceed by more than 35%, maybe the system should stop.
- Identification of duplicate decisions is another important control that can prevent problems in the operational systems from causing serious damage.
- Once the analytical controls are in place, they also need to be monitored to ensure they remain relevant to the changing business scenarios.
- Additional controls may be needed to track the changes in controls.

How sophisticated the controls need to be and how much influence they need to have on automated business decisions will vary from business application to business application. Financial transactions certainly would need a higher level of sophistication versus something like office supplies procurement. The bigger the impact of the decision strategy, the higher the focus on decision monitoring and control

Reporting and Action

Some controls will be in real time and will be measuring and reviewing the actual decisions against established thresholds and patterns. Most of the reporting and analysis of automated decisions is almost identical to the people responsible for designing, testing, and tuning new decision strategies. Same data is needed by both, but an additional layer of monitoring is needed to reduce the chance of costly mistakes, incompetence, and corruption. Identical data will be recorded in the analytical datamart for champion–challenger and tuning analysis and the audit datamart since, essentially, both are trying to understand what happened with automated decisions. Auditors need a fast-reacting environment to contain the potential damage.

The audit datamart, similar to most datamarts within the data warehouse environment, would need a reporting tool with good ad-hoc capability so auditors can investigate issues. All the best practices of reporting from the data warehouse have to be put in place on the audit datamart along with tight security and access schemes. A somewhat automated control creation mechanism has to exist to keep track of new controls (metrics and their thresholds) getting added and previous ones modified or removed. If a control has changed, there should be some audit trail of who changed and when and why.

PART

Implementation

CHAPTER 7

Analytics Adoption Roadmap

Now that we have established exactly where analytics fits in the Information Continuum, and we know what the prerequisites are and what is the range of problems that they can solve, we will use that knowledge to see how an analytics program (or center of excellence) should be structured, presented, planned, budgeted, and launched. Throughout this book the emphasis is on simplifying the use of analytics and its technical implementation by using data and tools already available. To convert an organization into an analytics-driven enterprise, analytics has to penetrate all aspects of the business. There are two ways to achieve this: top-down and bottom-up.

Top-down would mean massive upfront investment in a leading suite of tools and technology and highly skilled staff in technology with functional knowledge of the various business activities. This kind of analytics program can only come from the top executives and it needs a lot of personal commitment and effort to ensure all levels of management adopt analytics as a way of life to business improvement. This book is proposing an alternate bottom-up approach. The bedrock of the bottom-up approach to analytics adoption and keeping the delivery simple and easy to manage requires a short trip down history lane when data warehousing hit the corporate scene and how it was adopted.

LEARNING FROM SUCCESS OF DATA WAREHOUSING

There are six lessons that we learned from the success of data warehousing programs across organizations of all sizes. A data warehousing team, technology, and users working to build and use data warehouse systems for reporting and analysis is an essential component of IT departments of all sizes.

Lesson 1: Simplification
The concept of data warehousing as originally presented by Bill Inmon (1992) was very simple, and the simplicity of the concept lead to its

immediate and wide acceptance. The idea was applicable to all organizations and impacted all management layers. The use of data for reporting was straining operational systems and there was no single version of the truth readily available from an integrated source. To solve these two problems, Inmon recommended to take out all reportable data and move it into a data warehouse (Inmon, 1992). This frees up the operational hardware resources from reporting and data-crunching workload and allows IT to integrate the information from multiple systems while moving data into the warehouse. This was a simple and useful idea that appealed to all managers and business users.

Lesson 2: Quick Results

The adoption of data warehousing initially took a top-down approach (Watson, Gerard, Gonzalez, Haywood & Fenton, 1999) where large, funded projects were approved and massive data movement infrastructure and teams were set up, including hardware, a wide variety of tools, consulting firms, subject matter experts of operational systems, and specialized technical staff in data modeling, ETL and reporting, and database development. A lot of these projects became "analysis-paralysis" projects, as the analysis of an entire landscape of operational systems was overwhelming and took longer. Without that, it would've been impossible to guarantee a good integration of all systems. It took months and even years to integrate a couple of large operational systems with no value to the business until the entire integration was finished. That approach was soon abandoned since it didn't deliver results in a meaningful timeframe.

A datamart-based approach became popular where functional user-facing databases were developed integrating all relevant data for that function's reporting and analytical needs. Managers loved the datamart-based approach because it yielded quick results and solved their data availability issues. Some technical precautions were introduced under the data warehouse (DW) architecture and best practices were established to ensure the datamart-based approach was connected by a data integrity thread and not replicating the issues of the legacy reporting (Kimball et al., 2008). So datamart after datamart, over time a fully integrated data warehouse was built as a bottom-up approach. As business and functional managers got a handle on this datamart-based approach, they even started hiring their own datamart development teams that built specialized datamarts exactly according to their needs while pulling data from the data warehouse.

Lesson 3: Evangelize

Every manager and department needs reporting, and there was no need for IT or data warehouse teams to convince management that this was important to their performance as managers and as departments. The hurdle was already there, related to integrating all the relevant data for the metrics needed to run the business, and the data warehouse delivered a visually appealing presentation layer for reporting. Large organizations today have dozens of datamarts serving individual departments and newer ones are coming up every day. The more managers see, share, and adopt, the more they ask for in terms of data and better analytical capability. Datamarts are also available with a prebuilt analytical front-end as packaged software products in specialized areas like risk management, campaign management, anti-money laundering, etc.

Lesson 4: Efficient Data Acquisition

Data warehouse systems built robust capability in handling various forms of data coming from different systems at different schedules. This capability is called ETL—extract, transform, and load—but we will be using ETL as a noun referring to a capability of moving data between a source and a target and applying some data processing logic along the way. ETL is starting to become a vast field involving all aspects of data management and has become a small industry called data integration (Thoo, Friedman & Beyer, 2012). ETL teams got very good at linking to operational systems and have an existing integration with all operational systems and a mechanism established for receiving and sending data. Once this capability is in place, accessing data and serving the various data needs of IT and business teams becomes fairly efficient, removing one of the biggest obstacles in data analysis.

Lesson 5: Holistic View

The business requirements for the datamarts within the overall data warehousing program come from the functional owners of the datamarts—the users of datamarts. Once they see the capability of a datamart (may be within another department), they want one for themselves and provide the budget for building it. Data warehouse teams have business analysts who work with functional users to understand their data and analytical needs in the context of their business processes and operational systems. This results in the data warehouse team becoming the only group in the entire organization that has the complete functional and data perspective of the organization. This capability allows for business users to trust the data warehouse business specialists, and an exchange of ideas takes place for newer ideas on using data. These analysts are able to build better bridges with the business and internally exchange interesting ideas of information utilization helping the organization move up in the Information Continuum.

Lesson 6: Data Management

Data warehouse systems routinely deal with large volumes of data with millions of records processing regularly in increasingly smaller batch windows, storing terabytes of data, serving complex queries, and maintaining a high availability backup, archival, and disaster recovery setup. This gives them the capability to handle complex data problems. If a data warehouse engineering team has been around for a while and manages a multiterabyte-size environment, they are skilled in challenges of data. In addition to the technology infrastructure to support the data warehouse environment, they also manage the data dictionaries, data models, data lineage (how did the data reach the reporting display), and various other types of information like data-quality controls, data dependencies, transformation business rules, etc., typically called metadata. This capability allows them to consistently and reliably deliver quality datamarts to the business within the data warehouse program.

The success of the data warehouse is the proliferation of the datamart culture, driven and pushed by individual functional groups, allowing data warehouse teams to build a holistic technical, functional, and information landscape systematically over a longer period of time. That is why data warehousing was adopted and is extensively used in all aspects of the business. Analytics will have to follow a similar path and leverage the existing capabilities built within the data warehouse teams, since based on the Information Continuum, data warehousing is a prerequisite for analytics anyway, and therefore creating an analytics program office outside of the data warehouse team does not make sense.

LESSONS LEARNED

Six Reasons of Success of a Data Warehouse

1. Build the first pilot (datamart) through a technology-led initiative and demonstrate to key stakeholders as part of a bottom-up strategy for adoption.
2. Quick implementation and delivery so results can be demonstrated in a timely manner.
3. Roadshow the pilot to other functional groups so they start asking for their own reporting solutions (datamarts).
4. Ensure that enterprise data is accessible efficiently. Build a framework for a holistic view of the data as a by-product, building piece by piece with every datamart.
5. Ensure that holistic functional knowledge is built within the team fostering a better relationship with the business and exchange of ideas.
6. Build a robust data management skill around size, volume, and metadata management.

These six lessons learned assume that a robust and successful data warehouse program is in place. This is important because several of the first few levels in the Information Continuum are delivered through a data warehouse and analytics sits on top of it. If those levels are not there or are in a state of chaos, then undertaking analytics will be a costly affair with fragmented value and success. Datamarts being built and rolled out to demanding user communities in an efficient manner is a prerequisite for an organization to even start thinking about becoming analytics-driven. If your organization doesn't have the data warehouse capabilities as mentioned here, you can still undertake small analytics projects to illustrate the value and concept to the business. But once there is major business buy-in and support, do not proceed without making sure the maturity of the underlying levels in the Information Continuum are part of the scope.

THE PILOT

Business Problem

Analytics projects, pilots, or proof of concepts all suffer from the same constraint: the business may not know or understand what they want and IT wouldn't know what to build if there is no clear goal or scope. Chapter 8 is exactly designed to handle this problem. It doesn't matter if the business side or IT identifies a problem where analytics can be applied, because the problem can always be vetted and confirmed with the business owners and subject matter experts. It's a two-step process. A truly entrepreneurial approach is needed where a problem is identified that can be solved with applying analytics, and then rationalization or vetting is needed involving more subject matter expertise.

Chapter 3 uses examples from several industries to build a consistent theme around problems that can be easily solved with analytics. Important features of this theme are:

- Questions regarding some future event or activity. Would it be valuable to business if they could predict or forecast some future event or action and be ready to exploit that?
- Can a problem be broken down into a 1 or 0 problem? For any problem that can be described in the 1 and 0 format (Chapter 3 covers this in detail), historical data can be used to identify the patterns that predict 1 versus 0.
- A large collection of data that needs to be broken down into some kind of groupings based on similarities among them so meaningful analysis and actions can be carried out.
- Clarity of action to be taken (decision strategy) in case the future event can be predicted.

Some elements of these features have to be used to identify and build a business case. An analytics business case should be avoided if driven from hype, peer pressure, or vendor offerings unless the following are clearly understood:

- What is the analytics class of the problem being solved? Is it clustering, classification, forecasting, or decision optimization?
- What is the candidate problem statement to be solved? Open-ended discovery or exploration projects are risky, because if value is not uncovered quickly, the entire analytics initiative could be in jeopardy.

The discovery and exploration type of open-ended problems should be attempted once there is wider adoption and acceptability of analytics as a viable technology to improve business operations.

■   Is there a business buy-in to the problem and expected outcome? If a useful analytics model is produced, does the business understand how they would use the output?

In the absence of clarity on these questions, undertaking analytics solutions will not yield results. There may be some "informal" data exploration and subject matter discussions until a problem class is identified, a problem statement is crafted, and a decision strategy is understood. The Information Continuum levels preceding analytics models can help rationalize an analytics problem. If a general area of opportunity is identified but not a clear-cut problem definition, reporting, metrics, patterns, visualization, etc. can help structure the problem parameters leading to a formal problem statement.

Management Attention and Champion

It would be rare to find middle management coming to IT and asking for the initiation of a project that solves a particular problem using analytics. Even though an increasing number of universities are now introducing graduate-level degree programs in analytics and several programs are introducing the concept to MBA programs through introductory courses (Thibodeau, 2012), it will be a while before business managers will start initiating analytics projects. The lesson learned from data warehousing should be applied here, where a demonstration of something useful brings the managers across all functions of the organization to want their own datamart and reporting capability. The first datamart in a data warehouse program is always crucial and ends up being sales or finance focused. Finance in particular has a large appetite for data, so a lot of reporting delivered in the first iteration of a data warehouse is usually for finance. Besides, they need the most accurate and integrated information from all across the enterprise, so their needs are hardly met from departmental reporting.

Finance departments played a key role in promoting the use of integrated reporting and helped promote the datamart culture, which then departments adopted, and slowly data warehousing, datamarts, and business intelligence were part of the business lexicon. Analytics needs a similar champion or a natural starting point. The easiest could be a customer-facing CRM and marketing type of function of which the analytics needs are well established and various problem statements are well understood. If it can be demonstrated that some propensity models (predictive) or clustering can actually be carried out in a cost-effective way using existing data warehousing resources, that may be the easiest sell. Similarly, established problems in sales forecasting (using time-series analysis), pricing, or risk management are all good candidates as problem statements. The value is in delivering the solution cost effectively. Attempting a new and innovative use of analytics to identify a business problem, establish a problem statement acknowledged by business, and then deliver valuable results that are actionable is no easy task. So one approach would be to build confidence in analytics through delivering expensive solutions cost effectively, and therefore executives whose problems are being solved for which they know they don't have the budget will become champions and will be promoting analytics among their peers.

Another approach could be to find a data-savvy executive who is always asking for complex reports treading on statistics and patterns or trend analysis, and explain to him or her the four classes of analytics problems (descriptive, predictive, forecasting, and decision optimization). Then brainstorm some of his or her obstacles to see if any of these can be applied to build a model that can solve the most pressing issues. A few ideas will emerge; go back to the lower levels in the Information Continuum and do more analysis to see if a problem statement can be structured and bring it back to the executive explaining the inputs and outputs and the function of the proposed analytics model. This type of partnership is more valuable in establishing analytics as a tool for business foresight and success of the executive will propel the adoption of analytics. A word of caution is not to pick a problem for which the detailed and relevant data is not yet in the data warehouse. If the data is not in the data warehouse, it will be difficult to vet the idea in the absence of the lower levels of the Information Continuum applied to that data set and the implementation of the pilot will take longer because of the effort of source data analysis, cleansing, integrating, transforming, and loading, and then reporting and analyzing for metrics, performance variables, models, and strategies.

BUSINESS PROBLEM FOR ANALYTICS PILOT

Do not pick a problem for which the relevant data is not yet in the data warehouse!

The Project

Once a problem statement has been established and management attention or a data-savvy executive is willing to champion the pilot, focus on delivering the results quickly. The scope of the pilot, its candidate input variables, the analytics model, and its output and the use of that output for business decisions, all need to be carefully documented. There is no harm in bringing in a consulting firm or a product vendor into the pilot at this time. Help may be needed in using the open-source or in-database analytics capabilities. Training helps but expert resources in that particular tool are always more useful to have on the team. The deliverables will be along the following lines:

1. Problem statement.
2. Candidate variables as input into the analytics model.
3. Mapping of input variables to data warehouse data.
4. Design of a specialized analytics datamart.
5. Design of ETL to extract data from the data warehouse, transform and build variables, and load in analytics datamart.
6. Integrating and accessing the analytics datamart through the analytics software (data mining software that has the required analytics algorithms).
7. Identifying a large enough data sample to be used for the pilot.
8. Loading the analytics datamart.
9. Separating 90% of data for training and 10% for testing and validation.
10. Running and building training models.
11. Validating the results, adding more variables, and repeating steps 2–10 until good results are demonstrated by the model.
12. Feeding new data to the model and returning the output to business for actions.

During the pilot, it is unlikely that business will accept the model's output, create a decision strategy, and act upon it. At a minimum, they should see how the output can be used to benefit their operation. Decision strategies can be built and run as a simulation to even prove the actual return on investment of the project using data in retrospect (meaning already completed events treated as if they didn't happen and run them through the model for output and application of simulated actions). Compare the results of the retrospective simulation of decisions and with the actual decisions that were taken without analytics.

Existing Technology

To ensure the controlled scope and costs of the pilot, it is important to use the existing data warehouse hardware and software, including existing ETL and reporting tools. Do not invest in isolated, specialized, and standalone software or packages at this stage. Buy-in from management and adoption of analytics across wider business areas will automatically create the need and funding for such an investment, and the team would know how to use the specialized product to get the most out of its abilities. It is not uncommon to see cannons purchased to shoot flies (i.e., a large expensive software suite for a small problem) or acquisition of shelf-ware (software that sits on the shelf once the vendor finishes the project). There is no justifiable requirement for scheduling software, cleansing software, metadata repository, fancy front-end tools, and high-end analytics software at the early stages of analytics adoption.

There is a tendency among technologists to get the most well-known, leading, and powerful feature-rich software once the pilot scope is approved. Since the business has bought into the problem statement by now and expecting some benefit, IT approaches them and tells them that it cannot be done unless such and such software is purchased. Another scenario is where a vendor enters in with deep discounts of its software for the pilot with the assumption that a successful pilot will allow the management to secure the budget for the full price of the software to build a production-ready implementation. This has to be resisted by the business and technologists alike and rather use freely available tools like the R suite or in-database analytics software they may already have with SQL Server, Oracle, Sybase, or Teradata database environments housing the data warehouse. There may be a small price associated with turning on those in-database analytics (mostly based on data mining) features depending on the licensing terms, but that is the preferred toolset to be used for an analytics pilot.

Another tendency is to take all the relevant data out into another environment as a data dump and then try and build the variables, applying business rules and other requisite transformations using some combination of Excel, SQL, and other programming languages that the team may be familiar with. This can happen for two reasons. One is that the data warehousing team, too busy in managing the day-to-day activities of a data warehouse, may not want additional overhead of this R&D because they may not understand it, may not have the bandwidth to support it, and don't see much value from the pilot. They would prefer to just hand over the data and be done with it. The other reason could be that the project team, consulting firm, or vendor may find the controls and change management processes of a data warehouse environment to be too restraining and limiting and would prefer more freedom over the creation of tables, adding fudged data, dropping and adding columns and indexes at will, and introducing their own tools, techniques, and approaches into the development process. There may be some merit to both reasons. However, the temptation of creating an island of a specific data set for the pilot should be avoided. The more the pilot is integrated with the data warehouse, the better its ability to innovate in terms of variables, modeling techniques, and running simulations to validate decision strategies. An isolated implementation will struggle to scale up in terms of tapping into more data sources, building a robust data management environment, and providing adequate audit, control, and governance capabilities. It tends to become a one-person show solving a specific business problem, thereby limiting the organizational benefit from the successful implementation.

Existing Skills
Similar to the existing technology, the existing skills available within a robust data warehouse team should be leveraged. The skills similarity between a data

warehouse environment and an analytics solution has tremendous value and overlap.

Source System Analysis

The data warehouse taps into a lot of source systems, pulls data out, cleans it, integrates it, and loads into the warehouse. This requires subject matter expertise in source system technology, data layout, business processes linked to the data creation, and data definitions down to each field and its possible values. This skill is extremely useful in defining the variables as input into the analytics software. The candidate variables need detailed definitions in terms of the business rules, grain, possible values, and transformation to make them more useful for analytics. For example, a continuous variable like date of birth will have to be made discrete into age bands for more effective use in the analytics software. A source system analyst well versed in that data and its business process is best equipped to help build the age bands useful for the business. For example, in healthcare insurance systems, age 0 is very important from a reporting and analysis perspective and may need its own band. Data warehouse teams typically have field-level domain knowledge, otherwise they know the domain experts who can help. Navigating the organization to find the people with relevant knowledge of data and business processes quickly can avoid costly discoveries and realizations later in implementation.

Data Modeling

The data modeling capability within the data warehousing team is usually fairly sophisticated. If the data warehouse has been in production for more than five years and has four to six datamarts, the data modelers supporting the environment are well versed in complex data modeling challenges. They know how to work with very high-level requirements and develop the understanding of different data entities and their interrelationships. They are also well versed in abstract designs to handle the variations in data structures coming from source systems. That skill is needed to build an analytics datamart. There are no specific modeling techniques or design guidelines for an analytics datamart. Any modeling methodology can be employed, but the purpose of the datamart and the requirements are usually very clear by the time a pilot gets to this stage. This allows for the modelers to design an analytics datamart without acquiring any new resources (software, training, or consultancy).

ETL Design, Development, and Execution

ETL stands for extract, transform, and load and this book uses ETL as a noun, although data integration is starting to become an acceptable term to refer to all types of data manipulation and management (Thoo et al., 2012). ETL, therefore, refers to the effort required to move data from one place to another.

It involves some software, some design principles and methodologies, some hardware, and some trained and skilled people. The ETL design and development skill should be available within the data warehouse team and it should be leveraged along with the methodology and best practices.

The approach for analytics solutions presented in this book leverages the existing ETL skill set and breaks down most data movement–related pieces of an analytics solution into tasks that can be easily served through an existing ETL capability. These tasks include:

- Data for creation of variables needed for the analytics model comes from the data warehouse. Multiple sets of variables created by writing independent ETL programs built by different ETL resources allow for speed and efficiency.
- Loading the variables into a target data model for the analytics datamart.
- Running the model through a large data set and bringing the results back into the datamart.
- Building an audit control mechanism for data validation and integrity.
- Packaged analytics suites implement all of this data movement in easy-to-use custom and proprietary tools, but then there is the cost of acquiring those tools.
- Decision strategies can also be implemented using ETL tools, as shown in Chapter 5.
- Multiple decision strategies can be run and the champion–challenger process can easily be handled within the ETL tool.

ETL (software, hardware, process, and skill) is already available in any data warehouse environment. If the task is broken down into specific data movement activities, existing skilled ETL resources should be able to design, develop, and run the data within an analytics solution. More sophisticated decision strategies or data transformations (especially unstructured data like videos, voicemails, tweets, etc.) will require custom development or specialized software, but that type of complexity should be avoided for the pilot to minimize the risk of schedule and budget overruns.

Metadata Management and Data Governance

Within the data warehouse team, there is some mechanism of tracking the metadata. It is not required to have a repository to be installed with all the metadata integrated in one place (it would be nice though!), but some metadata management process should be in place for a data warehouse that serves multiple departmental datamarts. Even if the process requires metadata to be stored and maintained in Excel, it is acceptable for the pilot. As the adoption of analytics grows, even if the data warehouse couldn't make a business case for a metadata budget, analytics will. Source target mapping, a data dictionary, a logical data model, and some business rules around transformations are all usually kept and maintained somewhere within the data warehouse infrastructure (or with experienced resources). There are staff members who know how to build and manage this metadata. An analytics pilot needs to leverage that and just add its unique flavor of source target mapping, business rules, data model, and dictionary.

Similarly, there may be a data governance process with its two typical flavors: the design and documentation of the data warehouse, and the load statistics to detect anomalies and process load errors. If the governance process does not exist even informally, it has to be introduced as part of the analytics project. If the governance process exists, then it should be leveraged to control the design and definition of the data structure and track load and processing statistics and model and decision strategy versions in production.

Job Scheduling and Error Handling

Since we have broken the analytics project into smaller data movement and processing tasks, the scheduling of these tasks and error handling will require management. It is recommended that people responsible for managing this for the data warehouse should also be used for the analytics project. They may have existing templates and questionnaires that should be followed for the analytics project and they would do the rest in terms of connectivity, scheduling, dependencies of tasks, and error handling and standard support protocols when jobs crash.

Database Management and Query Tuning

If the variable building process, model training or build process, and the output analysis parts of the analytics project are not performing at an acceptable level, the same people who are responsible for these tasks in the data warehouse should be used to help with the analytics project. Similarly, table creation, indexes, constraints, more disk space, and all such rudimentary database administration tasks should be delegated to the team responsible for these functions for the data warehouse.

Reporting and Analysis

Once variables are created and loaded, the model is built and tested, and new transactions are scored or predicted from the model, it is all stored within the analytics datamart. Decision strategies and decisions made regarding them are also stored as data within the analytics datamart. Reporting is now needed to review the performance of the model, and the existing reporting resources used by the data warehouse should suffice to write analysis reports to review the performance of the model. This should be used for a roadshow—even some live and ad-hoc reports demonstrating the power of analytics should be used to increase the "wow" factor.

Results, Roadshow, and Case for Wider Adoption

It is very complicated to explain the working or output of an analytics solution to business and technology executives. If you tell them with great excitement that your prediction model is performing at 75% accuracy, they may not understand what that means. The explanation of results is crucial for the success of the pilot, not only for the stakeholders whose problem was being solved, but also for other executives to understand the possibilities. A walk-through of the current business operation without any foresight should be the starting point and then an introduction of foresight and ensuing actions should be explained.

Problem Statement

The first thing to explain while sharing the results of the pilot is the problem statement. Let's take the example of a warranty department of an automobile manufacturing organization. The problem statement would be: What is the probability that the next car off the assembly line will claim warranty in the next 12 months? Even if the project sponsor is the warranty department's senior manager, operations managers, engineers, and suppliers also have an interest in this statement. Keeping the problem statement simple and effective is absolutely critical when sharing the results. It should be obvious that if the problem statement prior to pilot development is not this crisp and effective, the results cannot be easily explained. Therefore, the simplicity and business relevance of the problem statement has to be emphasized in the early stages of the project, otherwise it will come back and haunt a team when explaining results.

This reasoning is why Part 1 of this book is restraining from adopting analytics projects with an open-ended statement like, "Let's find some interesting patterns in the data." If you don't know what you are looking for, it is very difficult to explain to someone what you have found. Also, whether something "interesting" has been found or not is not easily known unless a problem statement is looking for something specific. If it is found, then the model is ready, otherwise go back to the drawing board and build more variables or use a larger sample.

Champion management or a data-savvy executive's interest, support, and acceptance of results-as-value will allow other departments, managers, and business units to also understand and relate to the value from an analytics adoption. The value of an analytics pilot project is not just for the department of which the problem is solved, but for everyone to understand and appreciate.

Data and Value

The next step is to convert the historical data used in the pilot into a simple data profile, like the following, and present it. For example:

- How many years of total production data was used?
- How many production vehicles' data was used (the 0 records)?
- How many warranty claims were used (the 1 records)?
- What was the total dollar value of those claims used in the model building?
- What was the average time period from the manufactured date when the claims were paid?
- Present a list of sample variables used in the model building.

After presenting the profile, explain the 90/10 rule where 90% of the data was used to discern patterns and identify the variables that influence the claim output and the degree of influence they exert on an output. There is no need to explain the variables and their exact discriminatory value (degree of influence); treat that as a black-box instead. Then share the test results on the 10% where the model predicted the warranty claim correctly and where it predicted something incorrectly. This is called a confusion matrix. Let's build one (see Table 7.1).

This is a very simplistic and management-focused explanation of the predictive model. Actual testing and validation of an analytics model is a little more complicated, but relies on the accurate prediction of the test sample for which outcomes are already known. The value of the model is that it can

Table 7.1 Sample Confusion Matrix

Record Explanation	Counts	%	Comments
Total cars produced used in the sample	100,000	100	Total set
Total cars with claims in first one month	6,000	6	The 1 records
Total cars with no claims in the first 12 months	94,000	94	The 0 records
Actual claims that were predicted as claims	4,900	81.67	Correctly predicted; computed as (4,900/6,000) × 100. This is the performance of the model.
Actual claims that were not predicted as claims	1,100	18.33	Incorrectly predicted; computed as (1,100/6,000) × 100. This is the error rate of the model.
No claim records predicted as no claims	86,000	91.49	Correctly predicted; computed as (86,000/94,000) × 100. This is the performance of the model.
No claims predicted as claims	8,000	8.51	Incorrectly predicted; computed as (8,000/94,000) × 100. This is the error rate of the model.

accurately predict 82% of the time if a car is going to claim warranty in the first 12 months. How that translates into a dollar value is a complicated question. The model is not telling what could be wrong with the engine, parts, or finish of the car; it is simply showing the probability of a claim is high, and presenting it in this summary form makes it easy to follow.

Roadshow

The results should be shared with engineering and assembly plant staff, vendors and suppliers, purchasing teams, other entities within the supply chain, etc. It should be shared with the upper and middle management and workforce supervisors along with experienced operations staff and subject matter experts. All departments that are dealing with manufacturing and warranty should review the results. The results should be simple and straight-forward and presented in a concise and interesting manner that everyone can follow.

Wider Adoption

The hope is that the pilot will trigger interesting additional questions and explanations. Engineering would definitely want to know the discriminatory power of variables, therefore, they will dig deeper and may introduce more data to further analyze the cause, requiring a more advanced tool or an additional project with their internal engineering data. The finance teams may ask for more financial impact and detailed cost of red tagging the cars and not shipping. Marketing may come in and analyze the customer perspective on the warranty claims to assess the customer relationship impact. Each of these can convert into additional projects and more application of clustering, classification, and optimization on the data. The data warehouse would need to grow larger and faster and the ETL will really have to beef up so a business case for a comprehensive analytics suite can be made.

The decision strategies as to what should be done if the warranty claim probability is high, is something that an analytics team will have to help business evolve into. The business in the past did not have this probability so they never designed an action around it. Undertaking a decision strategy at the pilot stage will be very difficult until the organizational or business owners themselves figure out how they want to react to the insight that analytics provides. It may not be a bad idea to suggest some strategies so the concept and approach are understood.

Requirements Gathering for Analytics Projects

This chapter addresses the challenges of gathering requirements in an analytics project. This is challenging since a business has probably never seen anything like it and might not know exactly what they want to accomplish. In Chapter 8 we relied on a pilot project to make a case for an adoption of analytics; this chapter shows how to convert that into requirements for an analytics project.

PURPOSE OF REQUIREMENTS

The purpose and significance of requirements in traditional IT systems development is many fold. For example, requirements are used:

1. To act as a bridge between IT and business communication.
2. To provide information on what exactly the system should do.
3. To give users the ability to test the system against expectations.
4. To determine the degree of success and failure for IT as perceived by the business.

Will this be true for analytics systems as well? What exactly is the role and structure of requirements in analytics solutions and how do we adjust the current requirements gathering practices covered in this chapter?

REQUIREMENTS: HISTORICAL PERSPECTIVE

To illustrate the unique challenge of requirements gathering, analyzing, and formalizing in analytics solutions, it is important to take a quick trip down history lane to see how requirements management has emerged and formalized. The establishment of computing as a mainstream tool for useful applications can easily be attributed to Enigma, a cipher machine used to encrypt communication used by several government and military institutions across Europe and North America. During the Second World War, German forces were widely using Enigma to encode their communication and Allied forces countered that by developing Ultra, which was a code-breaking machine (Lewin, 2001). Both these computing machines were calculation engines and contributed toward establishing computing as a dominant tool in solving complex equations. The use of computing by military and governments remained confined to areas where large and complex calculations were required. The use of computing machines allowed for faster computations and lower error rate than humans.

These computing programs were actually software in a very basic form compared with software packages of today and, therefore, the requirements were the mathematical equations behind the computation programmed into the machines. There was no distinction between business and IT in those days. Scientists were primary users of computing machines, and they built the machines' hardware, designed and wrote the software, and ran the system (hardware plus software) to solve computation problems. No formal requirements were necessary. As computing matured and moved into the commercial arena, the separation of IT and business started to emerge. Financial firms, government and military agencies, and their suppliers started purchasing specialized computers from vendors. Now, the users of the system were separated from the builders of the system. This created a translation gap between the problem and the solution.

ADP was among the first companies to adopt computing to generate payroll and print checks. ADP bought specialized machines that would compute payroll, taxes, benefits, etc., and then print checks for their clients (ADP Corporation, 2012). ADP was neither IT (building the system) nor business (ADP clients specified what they needed). This analogy is important to understand the role or purpose of requirements. ADP acted as the translator between what their clients needed and what they wanted the computer manufacturer to build for them. They owned, analyzed, developed, and published the requirements and then managed the operations built against the requirements.

Since those days, computing has evolved from calculations to process automation leading into data analysis and analytics. How has the role of requirements changed through this evolution? Figure 8.1 shows the evolution of computing from calculation based to the analytics of today.

Calculations

The Enigma and Ultra machines belonged in the innermost section of Figure 8.1—that of calculations. Various other implementations used for scientific computation also belong in this space. Systems belonging to this section are usually built for massive computations that cannot be undertaken by people in a meaningful timeframe and with a reduction of errors. In the 1940s and 1950s, with these calculation systems, businesses (users of the system) and IT (builders of the system) were the same group of people, and therefore no formal requirements were needed. People well versed in mathematics, electronics engineering, and computer science worked side by side as a team to implement these systems.

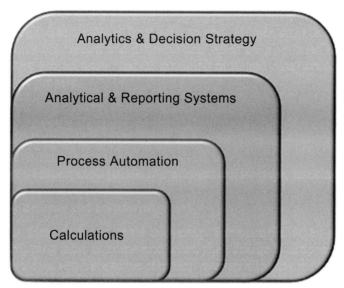

FIGURE 8.1
Evolution of computing.

Even today, there is plenty of evidence of this type of approach for solving specific problems in meteorology, nuclear science, mining, and metallurgy, and increasingly in Big Data where purpose-built machines have to be engineered using commodity components. In such scenarios, hardware, software, and domain experts work side by side to build such machines, and therefore formal requirements are not needed to bridge a translation and expectations gap. In places where the problem domain is specialized, such as government's census bureau, central banks, departments dealing with statistics, etc., they have to deal with IT and business as two separate entities, but the problem domain is simple since the application of computing is limited to calculations, not to carry out all of their functional activities.

The four points that made up the purpose of requirements are:

1. Requirements have no need to act as a bridge between IT and business communication since both roles are with the same people.
2. There is no need to formally document what the system should do since there is little ambiguity within team members about the purpose of the system. All participants are trained in the same discipline.
3. The testing is also carried out by the builders since output is the result of the calculations.
4. The builders and the users both understand the degree of success from the project looking at the results.

Process Automation

The ADP solution belongs in process automation space and dominated the computing and IT industry from the 1960s to the 1980s. In the 1990s, this was taken over by ERP (enterprise resource planning) systems. In process automation, the requirements capture the current business processes and any changes desired by the business. The requirements capture, analysis, rationalization, and overall management of requirements and the skill to carry out these functions are well established and requirements documentation methodologies are fairly mature. Therefore, requirements gathering and analysis became a formal discipline in process automation that relied heavily on documenting how things work today. The requirements process would document the existing operations and activities and then design a system to automate that process. Various requirements gathering tools and techniques have been developed and are widely used. In fact, whenever the discipline of requirements gathering and analysis is reviewed today, it refers to this category of requirements within process automation space.

Analytical and Reporting Systems

Process automation uses computing to automate all paper and information flow. As it accomplishes that, it creates data as a by-product. Analytical and reporting systems rely on this data to build a different kind of system that requires a different kind of requirements gathering approach. In the case of process automation, the knowledge and expertise as well as the obstacles of the existing manual processes are well understood by the business. That knowledge becomes the requirements of what the process automation system should do. For analytical and reporting systems, the challenge is that there is no manual or legacy precedent that can be used as a set of requirements. Therefore, the requirements gathering process actually becomes the requirements extraction process, where the business is shown what is possible and they determine what would make the most sense for their organization. This is why it is common to see requirements change dramatically once a data warehouse system is rolled out. Immediately, the business starts to get ideas on what can be done. The requirements extraction process therefore requires considerable knowledge in

- The business domain.
- Specific business processes under consideration.
- Data and its interrelationships within those business processes.

The effective requirements extraction process requires people well versed in the business and the wider industry, as well as data used within that industry. For data warehousing, typically a set of legacy reports are used to initiate the design and a datamart implementation. Once the ease and effectiveness of the reporting from a data warehouse is evident, the users (usually the savvy ones first) come back with more sophisticated requirements. This is the only way to move up the Information Continuum to get more out of the information asset of an organization.

The requirements gathering process for such systems requires questionnaires and explanations of advanced concepts in historical data analysis, metrics and their thresholds, time-period comparisons, snapshots, trends, and the need for cross-departmental or cross–subject area information content. These concepts are explained to users within the context of their business process and the data they deal with in their normal execution of duties. Once they understand the possibilities, they are better able to articulate their requirements. This approach puts pressure on the analysts tasked with extracting the requirements since they have to be masters of the nomenclature, holistic business perspective, existing systems and processes, as well as data within those systems. Additionally, they should also have the prerequisite knowledge of surrounding business processes that act as input or receive the output of the process under review.

Analytics and Decision Strategy

The analytics and decision strategy requirements are more complicated than reporting systems. This is for several reasons:

1. Overall, the leap for analytics into mainstream is in its infancy.
2. Analytics is not part of the mainstream manual processes so no precedent exists.
3. Use of analytics is limited to small specific problem areas.
4. The application of analytics cannot be limited to one business process at a time; it tends to look at a group of business processes as a whole.
5. The data needed for a useful analytics solution is never known upfront and a lot of domain knowledge and trial and error is required to identify the data that may be useful.
6. Analytics-driven strategies are nonexistent unless the analytics model is in place.
7. Benefits of analytics is dependent on various factors, including the creative strategy design and the effectiveness of the model for those strategies.

Based on these reasons, extracting the requirements is more challenging because skilled personnel in the business domain, integrated business processes, and understanding of data and areas of opportunity are not easily available, if at all. If the requirements extractors are not available, and since we already established that a typical requirements gathering approach is not an option, we are left with no requirements for IT to build an analytics solution.

REQUIREMENTS EXTRACTION

To surmount this challenge of requirements, we will work with the analytics requirement strategy that is covered here. In Chapter 2, there was a hierarchy of information needs where the use of data was shown to move toward higher business values and overall benefit to the organization. The requirements extraction for analytics relies heavily on that hierarchy through a series of questions, such as:

- What kind of data do you search for? What is the typical business reason for that search? What should be done if more than one search result is sent back?
- What kind of summaries and counts matter to your business functions? What do you do with those counts? Do you compare them to some other numbers? What happens if the numbers are too low or too high?
- Who reviews these reports (existing and available reports)? What are they looking for in these reports? Do they change the filters, sorts, and groupings?
- What kind of summary and aggregate reports or historical snapshot reports are used? Where does the data come from? How come one data type is reviewed like this but not another data type? How much data is retained historically? Is it purged or archived afterwards?
- What metrics matter to the business? How are the metrics related to business objectives? What is the calculation that goes into the metrics? Does everyone agree to these metrics and their definitions?
- What about users outside of the department who use these reports—who are they and why do they use these reports? How do the reports impact them?
- Are there any dashboards in place? Are there any geographic information systems (GISs) or other type of analytical applications in place?

It is evident that these questions follow the Information Continuum hierarchy and levels, and provide analysts with a structured approach to familiarize themselves with the levels before the analytics models. This is a line of questioning that allows an analytics business analyst to understand the following:

- Source systems involved.
- Business processes involved.
- User and business sophistication.
- Important data entities. A *data entity* is a loose term referring to a certain data type, such as a customer is a data entity and name, address, and age are the attributes. Order is another data entity and order number, order date, ship date, and order status are the attributes.

Most importantly, this line of questioning establishes where the users fall on the Information Continuum. If they do not reach the prerequisite level as

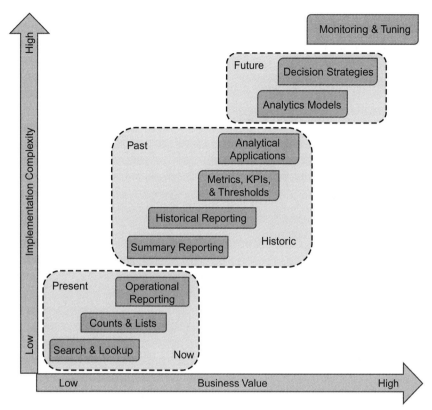

FIGURE 8.2
Information Continuum.

shown in Figure 8.2, then the business may not be ready for an analytics project, or the challenges or risk of a project's success increase significantly. This is where you may end up with a project that spends money and builds a lot of learning but the business may not benefit as much as it could if the business division, department, or group was at a higher level of sophistication on the Information Continuum. Sometimes this is good for IT personnel, but not very good for business, as the unimpressive results may leave a bad taste for analytics all together and it could be a while before a renewed interest in analytics is developed.

Problem Statement and Goal

Once it is established that the business is ready for analytics and has been dealing with historical data, metrics-based decisions, and has some idea on trends of the metrics year after year, the business analyst takes the lead in formulating a problem statement (see Chapter 3 on using analytics). The business analyst responsible for gathering requirements for an analytics project is expected to have sound understanding of the four analytics techniques to a point where their application can solve a particular business problem. The business analyst can explain these concepts in the context of the business process and ask questions, such as:

- What is the metric you use to measure the effectiveness of your operations?
- Is there a way to forecast that metric for future periods? How would you use that information?
- What if a certain value of the metric can be predicted? How valuable is that?

Similar to the metric, the same line of questioning can be performed on some of the heavily used reports. It is important to understand what the reports are used for, how the information in a report impacts the day-to-day operations, and what pieces of data in a report have an impact to the business because of their spikes and dips. Business analysts in the reporting and analytical systems usually stop after collecting the report layout and definitions and convert that into requirements against which data warehouse systems are built and run, but rarely do they actually try and understand how the changing information in the reports impacts the business.

The requirements extraction process starts with that analysis to identify important data and then formulate a problem statement using any of the four methods of analytics to see if the business can make use of that analytics output. This method is less science and more art and innovation, and this process of business analysts interacting with business users to extract an innovative use of analytics from their existing known data is an important process to master.

Newer kinds of data like tweets, cell phone–to–cell tower signals, RFID tag readings, social media interactions, location check-ins, likes and shares, etc. pose a new problem for requirements extraction because of the sheer volume and nature of its use. The use of that data has to follow the same Information Continuum, but the pace needs to be weeks rather than years in the current scenario. Some may argue that this new data type should enter directly at the higher levels of the Information Continuum, but that may result in isolated success depending on the creativity of the user. There are two reasons why this newer form of data (also dubbed Big Data; see Chapter 11 for more detail) has to follow the Information Continuum:

- The organization as a whole can learn to benefit from using this data if it follows an evolutionary path. This is similar to how customer-complaint data was not important to purchasing or wasn't even available three decades ago in a form that could be used. When that data was finally brought in to the fold and followed the evolution along the Information

Continuum, various other existing data pieces realized the value of using the additional information, and purchasing started to look at defects as another factor in selecting vendors and bids. So, intermixing of existing data with Big Data can only happen if an evolutionary path is followed.

■ If an evolutionary path is not followed, the benefits of Big Data will be isolated and sporadic where the individual brilliance of people will yield results but the wider benefits will be limited.

Therefore, the existing data has to fit into analytics using a process that includes some science (existing reports and metrics, and existing use of that data) and some art (where can the four methods of analytics help the business). Nowhere should analytics be taken as a brute-force approach and run on data, because then there is no guarantee what you will find and, whatever you do find, how to use it for business value. Granted that with Big Data, the speed of traversing the Information Continuum through the evolutionary process has to be very fast and sometimes data discovery–type approaches come in handy. But the data discovery really breaks down into one or more of the following options:

1. Geo-spatial analysis of the data.
2. Visualization of the data, such as in 3D or other forms of display within Euclidean geometry.
3. Data profiling (usually showing the mean, mode, frequency, minimum, maximum, distinct value, etc.) of fields.
4. Some basic correlations between variables that a user identifies.

Data discovery should be used to speed up the process of moving up the Information Continuum, but should not be treated as an end-all for identifying the data that matters before starting to apply analytics on it. The requirements extraction process should be able to identify the important variables through this type of interaction and review with the business.

Working Example from Consumer Credit

A small bank has a loan portfolio that looks like the following:

2,000 auto loan accounts	1,800 unique customers
$30,000,000 total loan originally given out	$23,000,000 still outstanding (being repaid monthly)
$750,000 in default	40 accounts with 40 unique customers in default
$1,500,000 in 30- to 90-day delinquency	$450,000 collected last month from default accounts
$1,000,000 total profit in the last 12 months	$150,000 written off as loss in the last 12 months

The branch loan officers who sold the loans, the risk manager, the collection supervisor, the finance department, and executive management all consume this data on a monthly basis at any lending institution. Various reports are generated that provide pieces of this information to the relevant users. The branch loan officers who sold the loans are interested to see which of their sold loans are in trouble. The finance team and executive management are more concerned with liquidity, bank's capital adequacy, provisioning of loss amount, and overall impact to the bottom line. The risk managers constantly review the reports, data slicing different segments and variations to see how to better structure the lending policy. Collections goes after the people who owe money and tries to maximize the collection amount against the collection effort. The Information Continuum looks like the following with examples on the same data presented above:

Search and look-up	Ability to look up customers to call for payment reminders.
Counts and summaries	How many accounts in 30-day delinquency are broken up by zip code? What is the total amount stuck in 90-day delinquency for provisioning purposes?
Standardized reporting	Capital adequacy and portfolio summary reports are typical standard reports shared with management and even regulators in some cases.
Summary and aggregate reporting	What are the total amounts of up-to-date loans; 30-, 60-, and 90-day delinquencys; write-off with their total borrowing; and provision and potential loss amounts? Another aggregate report can be related to the cost of funding against lending, interest income, and losses to determine profitability.
Historical snapshot reporting	Comparison of accounts in 60-day delinquency in current month to see how they were doing at the end of last month. Which ones slipped further away, which ones are recovering, and which ones are stuck?
Metrics, trends, and thresholds	What is the typical delinquency migration to see how many accounts will move from 90-day delinquency to total default? If that number is increasing, then the collections team needs to step up.
Analytical reporting	What is the overall portfolio situation with profit, loss, interest, lending, and their trending over the last year?

Once the business analyst has a handle on the above, proposing analytics problem statements becomes easier. Users who are savvy with this type of analysis through the Information Continuum get excited when presented with a question, such as: How would you like analytics to predict which accounts will self-heal back from 60-day delinquency to nondelinquent? This analytics question may not be obvious to someone who has not evolved through the Information Continuum with this data. But someone understanding this entire hierarchy will ask all sorts of additional questions, such as:

- Can you forecast what will be the losses in the next month (forecast method)?
- Can you predict which accounts are going to go bad (predictive method)?

- What is the probability that certain accounts in collection will actually pay-up (predictive method)?
- How should new loans be priced based on the current cost of funds, losses, provisions, and competitive market situation (decision optimization method)?

Chapter 9 explains what kind of skills and training are necessary for a business analyst to encourage and explain this type of thinking once the data along the Information Continuum is understood. The more the user community gets better at this, the better the value from analytics investment to all parts of the organization. Within banking, risk managers are usually well versed in analytics, but sales people in branches and collection teams usually are not. That is where IT can help bring analytics to the front of operations and create knowledge workers out of sales and collections staff. Once they get on to this line of thinking, the analytics solution team will have no shortage of requirements coming from all over the business. We have already seen this trend with data warehousing and the wide adoption of reporting tools and datamarts across departments and thousands of users.

Data Requirements

Once the problem statement is clearly identified as part of the requirements extraction process, it should have three components: the analytics method to be used, the data domain or the problem domain, and the resulting variable that needs to be predicted, clustered, forecasted, or optimized. The data requirements are a critical component of the analytics project because they keep users from blindly running around and hoping to get something of value. The existence of data is tied to some business process that could be generated from an internal or external system. If you are trying to forecast the sales for the next quarter, you may be wasting your time if you are looking into the HR system. So when the business starts to understand the four analytics methods and appreciate what analytics applications can do to run and manage the business, there is a lot of excitement and anticipation.

However, if you ask users to predict the possibility of a loan default and what data they should be using, you have the business in a quandary because they wouldn't know. On the other hand, if someone not well versed in the data and problem domain tries to identify the data that may be relevant, he or she can end up acquiring data sets that should not be combined, mean different things, or have no relevance to the defaults. So how does a business analyst become well versed and what exactly are these data requirements that should be gathered? The approach for data requirements will be similar to the one used for the problem statement. Business analysts will have to do some work on their own and then present the findings to the business and explain what they are trying to do. The business in turn will guide and help pinpoint the data requirements. It is important to use this approach because the following questions are very important to put meaningful boundaries on the project, otherwise it can become a science experiment that may not meet expectations.

1. How much historical data is kept around? What is the relevance of data that is more than three or seven years old?
2. How do you ensure a good spread of data is identified? Data that is skewed toward a dip or spike in the business process may have to be normalized to keep the analytics model from skewing its results.
3. How many source systems are relevant? This ensures time is not wasted on systems that may have only peripheral importance.
4. What grain of the data is most relevant for the problem statement? For example, is the prediction of defect on a single product or on the lot (group of same products produced as part of the same batch like pharmaceutical tablets)? Whatever the grain, the data will have to be built at that level.
5. What is the total set of tables and columns from various systems that should be in scope?

This type of information will allow the project to manage the data-quality issues, data transformation effort, data profiling and normalization efforts, and understanding the dependencies across various data elements. In large organizations, an analytics project may be looking at hundreds of tables and thousands of columns; the method of elimination as to which variables really have the weight to influence the prediction is a slow process and could take weeks and months. This is one of the serious limitations of the current analytics projects undertaken in specialized areas like risk management and direct marketing, where the analyst who knows the data from prior experience will go after the data that matters and will not waste time identifying other potential interesting variables. However, that approach only works when you have super-analytics professionals (highly paid) and well-defined problem statements and analytics methods. The purpose of this book is to take analytics to all parts of an organization, and those newer parts may not know readily which data elements matter the most.

Another issue with the current practice of creating a data silo for analytics and only using the well-known variables (e.g., age, income, gender, marital status, etc. for customer analytics) is that the Big data, tweets, likes and check-ins, geo-locators, uploaded pictures, recommendations, Angry Birds or Farmville scores, etc., cannot fit into those predefined variables and ensuing analytics models. On the other hand, bouncing around endlessly in Big Data variables can take forever without any guarantees of success. There is no scientific answer to this, as we have established that analytics is both science and an art form. A systematic approach is presented here to make sure the data requirements are deep enough to factor in the newer data types and have enough data, yet not overwhelm a practical delivery schedule of the analytics project. The business and IT analyst will have to work closely where the business analyst will do a lot of preparatory work and present findings to the business and seek their guidance to extract data requirements.

Data Requirements Methodology

Since the analytics approach is identified and agreed on and the data output of the analytics model is well understood, identifying the business processes relevant to that data is not that difficult. Here is a methodology to do this:

1. The business analyst should look at the people involved in the scope defining stage and jot down their positions in the organization chart.
2. List the organizational units involved in the analytics output setup as the scope.
3. Create a list of systems that the users within those organization units regularly use.
4. Create a high-level business process model of how the users, systems, and their organization units are connected (there is plenty of material available on business process modeling and it is usually well understood by business analysts).
5. Create a high-level list of data entities that are used in that business process model.
6. Map the high-level entities (also known as subject areas) to the systems and tables within the systems where that data lives.
7. Prepare a master list of all tables organized by the system.
8. Map the tables against the data already in the data warehouse.
9. If the gap between the tables list and the data warehouse is more than 20–30%, *stop*.
10. The project must not move forward until the data warehouse is able to source and bring in data from the corresponding tables based on their internal analysis, design, and implementation methodology.
11. If most of the data is in the data warehouse already, then the tables and columns combined with the metadata, such as source fields mapping, transformations, and business rules, all become the data requirements.

Step 9 is a guideline, as there may be situations where you may tap into the source system for missing data. The reason is that if the data not in the data warehouse is included in the requirement, then subsequent design and development steps will become more complicated and a sustainable solution will not be possible. Analytics solutions like the rest of the data warehouse are based on a continuous supply of data on a regular schedule that runs through the analytics model either as a real-time transaction or in batches, but either way, the extraction of data as a one-time use for an analytics project is a bad idea. If it is for a prototype, then Chapter 4 provides the methodology and process to build a prototype. Starting in Chapter 8, we cover a long-term and sustainable solution and the requirements for that are not for a one-time analytics project. This will require the data to be available on a regular schedule (or in real time) to run through the model, and therefore it is better to integrate the data into the data warehousing infrastructure to reap the benefits of the ETL, data architecture, metadata repository, scheduling tool, real-time integration with source systems, archiving and disaster recovery setup, etc.

Model and Decision Strategy Requirements

This and the next part of the requirements process should be driven by business once the concept is explained to them. Formal requirements extraction activity may not be needed. This part ties back into the scope and the problem statement. Let's take the example of a wireless phone service company looking to manage its customer churn. They want to build a predictive model that predicts the probability of a customer leaving the carrier. This has been established in the scope and problem statement and the relevant data fields for that problem have also been identified. The question becomes, what would the business do if they get such a probability against all of their existing customers? The purpose of analytics is not just to build a model and show the business that it is possible to predict the customers who may defect, the analytics solution and practice should also help the business figure out what to do with that information. Once this is clear to the business, they should be able to provide exact requirements as to what they would need from the model.

The decision strategy may take a little more explaining. As explained in Chapter 2, a decision strategy is basically a series of business decisions that factor in the output of an analytics model and provide actions to be undertaken by the wider organization. The business should be able to provide 1–3 strategies at a minimum to as many as 8–10 as a requirement. The team delivering the analytics solution will be responsible for building and implementing the strategies as well, and therefore the requirements gathering is also the responsibility of that team. In this particular case, business will get the probability of defection back from the analytics model against each customer (see Table 8.1).

Table 8.1 Probabilities of Defection			
Customer ID	**Probability of Defection**	**Customer ID**	**Probability of Defection**
1002356	67%	5012356	3%
1345098	64%	2928374	77%
8675629	13%	1029657	55%
9287462	23%	7565839	43%

The business should be able to decide what they want to do with customers with high probabilities and what high for them means (>50, >70, or >90, etc.). The answer should be a requirement that looks something like the following:

```
IF probability > 65 and < 80
AND
    IF Customer has been with the company more than 1 year
    AND IF has been paying more than $40 per month regularly
            THEN offer 20% discount for the next 3 months
            ELSE offer them 10% discount
    ELSE send a thank you letter that we value your business
IF probability > 80%
```

It is important to note here that the analytics model is only providing the probability, but what needs to be done with the probability is totally up to the business and is being captured as a requirement, however its implementation is not simple. The business may want to know a distribution of the probabilities to decide where they want to put a cutoff. In the preceding example, two cutoffs were used as an illustration: one was for everyone over 80% and the other was for customers with probabilities between 65% and 80%. What if based on actual results of the model, there aren't any customers in the 80% range and the 65% to 80% range has 1 million customers? The cost of providing discounts to 1 million customers for retention could be very high. Maybe the high number of potential defections is a question for a broader strategy shift in how the customers are provided services or a new product design may be necessary. These scenarios have to be presented to the business to get them fully engaged in analyzing the output so they can be prepared for the actions necessary to get the benefits of the analytics project. A high concentration of customers in fewer probability buckets may also prompt the business to ask for explanation as to what is causing so many people to be in the same bucket. Either the model needs tuning or there is a very interesting insight that business would like to understand. These requirements are a little more tricky to answer and significantly depend on the analytics tool used, but they should be captured so the solution team is ready and can test the model using some of these scenarios.

The requirements for the decision strategies are also very important because if the output of the model is well received by the business, they would want to act upon it. If at that time it turns out that strategy design, implementation, and deployment is another four- to six-month project, then the analytics team losses credibility. Some steps in the strategy like "AND IF has been paying more than $40 per month regularly" could actually be fairly complex to implement as other scenarios emerge about what regularly means. Is it trailing 12 months, year-to-date, or last calendar year? The answers to such questions can determine the complexity of calculating this segmentation and implementing the strategy.

Business Process Integration Requirements

The last part of the requirements gathering process is regarding operational integration of the decision strategies. Looking at the strategy requirements it is an action that can be automated and implemented in an operational system, such as the phone billing system in this case. A wireless carrier may have millions of customers, so it is unlikely that the decisions will be carried out in any manual form. But whether they will be carried out in a separate standalone analytics system or integrated with the existing operational system requires further probing and engaging the business more closely. The questions to be asked would be:

- Are the same customer service people going to implement the decision strategies?
- If this is implemented in the operational system, how often would you change the discount percentages and probability ranges?
- Are all the customers going to go through the analytics process every month or every quarter? (If yes, then you would be giving them a lot of discounts.)
- Should the analytics model run on select customers based on some event?
- What kind of controls should be put in place so no other customers get the discounts through clerical error, misunderstanding, or foul play?
- How would you like to review the results of the discounts once they have been applied to the customer accounts?

These requirements gathering questions should make it clear that the analytics solution is a complex implementation, because the answers to these questions can significantly change the complexity of the implementation. A specialized rule engine may have to be purchased or built integrated with the operational system that interacts through transactions, clear interfaces, and integration mechanisms. The entire overhead of changing the decision strategies, adjusting cutoffs, and modifying discount values, and tracking the decisions for auditing purposes, may need to be kept out of the operational system.

The business, once fully engaged in the analytics solution implementation, should be able to provide all of the preceding requirements. However, they may also get overwhelmed and start asking questions as to why this is being done in the first place. That is where a comparison of their current business process for reducing churn may have to be explained. A nonintelligence or data-driven churn reduction strategy may be costing them more and not providing the desired results in their current environment. In this case, the solution is complex, but business has to take ownership and control of the solution's moving parts because then they will have a wider visibility into where to focus their energies. Without the churn probability, they may have been offering discounts to customers who were not going to leave the business anyway. This level of engagement helps the business understand the concepts and provides the appropriate requirements under each category to allow the entire business function to become knowledge-driven and the staff to become knowledge workers.

CHAPTER 9

Analytics Implementation Methodology

This chapter provides an overview of the analytics project methodology that needs to be followed to enable the successful implementation of an analytics project. Consistent with the theme throughout this book, analytics projects are not a one-time activity, rather a new business lifestyle, and as more and more processes get innovative and customized to the customer, product, and operational needs, analytics-driven impetus into those processes will be a key requirement. Therefore, analytics projects have to be built and managed with a structured and methodical approach that is reliable and repeatable. The reliability is far more important for analytics since daily business decisions are made relying on analytics output.

Analytics projects, analytics technology, people, and the problem space are very different from traditional IT projects. The main differentiating factor is the definition of the problem. In traditional IT development or even within data warehousing, the problem space is defined by existing processes and existing analysis, metrics, and reports. That provides enough direction for solutions and newer technology applications to address existing issues and accommodate innovative requirements needed to improve processes. For example, new smartphone- and tablet-based applications that open up a field staff to get things done while on the road is a new business process needed within the same problem space.

In traditional development the problem statement either comes from an existing issue, from business innovation, or from IT where they demonstrate the possibilities of newer technology. If analytics solutions worked the same way—that is, that it's just a new technology for an existing problem space—the following would be some possible scenarios to initiate an analytics project:

- The business can come forward with a specific need of an analytics solution to their existing obstacle. This would be a traditional way of IT and business engagement. But this is rare apart from some well-established problem domains like direct marketing, customer profiling, risk management, supply-chain optimization, and financial trading.

- The analytics team, learning from a wide variety of problems presented in Chapter 3 or similar case studies and other industry publications, reaches out to business with ideas and tries to form a problem statement.
- The IT analytics and business subject matter experts play with the existing data and try to find a problem that can be solved using analytics methods.

All three are legitimate ways of initiating analytics projects, but an organization's approach for democratization of analytics will heavily influence how these scenarios play out.

CENTRALIZED VERSUS DECENTRALIZED

The first issue that needs to be tackled head-on is whether the analytics approach for any organization is centralized (top-down) or decentralized (bottom-up).

Centralized Approach

In a centralized strategy, all the analytics assets, such as the software, hardware, skilled resources, and other information assets like metadata, business relationships, and the project management capabilities, are all under one organization. This approach has a lot of benefits of knowledge sharing, consistency of implementation, consistency of support, visibility and clear lines of communication, and, above all that, reduced cost as hardware and software assets are shared across various projects. These benefits may be tempting and make a lot of sense, but then getting to this centralized state throws an almost insurmountable set of challenges related to business case and budget approvals.

The premise is that the hardware and software should be procured first; then the team hired and trained; and then the engagement, project management, and development methodology built so the overall architectural framework is in place before real projects start and deliver value. It is very difficult to find sponsoring executives willing to fight for these kinds of approvals and budgets. And it would be very tricky for an IT executive to have internal resources to pull this off on his or her own. So the centralized approach will remain stuck in business case and budgetary approval processes.

In addition, the lack of business insight and specialization in particular business areas that can most benefit from analytics can also become a problem with a centralized approach. Pitching various analytics-driven improvement ideas to business executives requires in-depth understanding of the underlying processes and data, and it is not possible for a centralized team to build that capability across the board. Therefore, pretty much the only available option is when business comes with a defined problem statement, the analytics team can deliver value. That limits the usefulness of a centralized team and investment.

Decentralized Approach

A decentralized approach would imply that each department or business unit is fully aware of their needs and, understanding how analytics would help, buys software, procures the hardware, and hires a team to build their solution. The team is owned by the business unit or department and is dedicated to building and improving the analytics-based decisions for the department. Consulting firms and specialists in that business domain may also get hired. This is the most common approach for analytics and marketing departments, risk management departments, and other specialized areas within various industries like manufacturing, healthcare, banking, and insurance that use this approach.

The main benefit of this approach is specialization in a business domain and adoption of industry-leading practices for well-defined problems. But all along the intent of this book has been to show how an entire organization can benefit from analytics, not just one or two specialized areas. This decentralized approach cannot be applied to the wider organization because that would imply each department has their own hardware, software, and specialized teams. They may benefit from a specialized team and software, but in untested problem areas and a new innovative use of analytics, this approach will not work because of lack specialists in the market and problem-specific software. Not to mention the cost overheads of every team managing their own technology and resources, as they may not be able to fully utilize the capability acquired.

A Hybrid Approach

It should be obvious that a hybrid approach is needed for democratization of analytics, and therefore the methodology being proposed is based on that conclusion. The strength of a centralized technical team, a single enterprise instance of analytics software deployment, and shared hardware infrastructure make a strong case for following a centralized approach. However, then business domain–specific analytics business analysts and analytics specialists closely aligned with the respective business units and departments are needed. A matrix approach with dotted-line reporting for these two roles (analytics analyst and analytics specialist) will allow the business units to understand the capabilities and possibilities, while the dual-reporting roles will work closely with the implementation teams to deliver solutions.

BUILDING ON THE DATA WAREHOUSE

The centralized parts of the hybrid approach dealing with technology and data processing skills are available within the data warehouse teams. Looking back to the Information Continuum, a robust data warehouse infrastructure

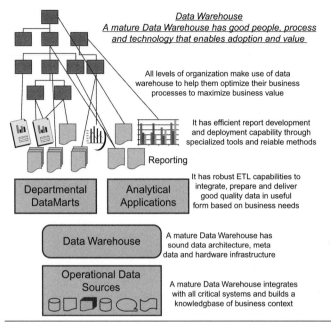

FIGURE 9.1
DW capabilities.

(including its hardware, software, and skilled teams) is a prerequisite before an organization jumps into this hybrid approach for analytics. A data warehouse has the following working pieces already in place:

- Knowledge of source systems, business processes, and data.
- Integration with various source systems.
- An integrated data structure with all the data loaded in one place.
- Business-specific datamarts where relevant data for each business unit or department is available.
- Dashboards, metrics, KPIs, and historical perspective of important data being used for decision making per business function.
- Hundreds of reports that have been designed, developed, tested, and run on a regular basis, so there is an overall good idea on what data is important.
- Alignment with power users across business functions who are the champions and promoters of increased and innovative information use for running the business.
- The reports are consumed by all parts of the organization, so there is a clear idea by department who is consuming what.

Figure 9.1 shows the capabilities of a good data warehouse program.

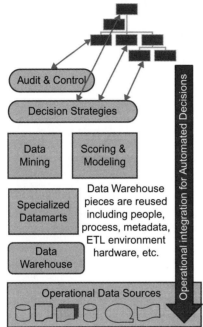

Analytics
Analytics should have simlar building blocks
to increase adoption, bring delivery costs down,
and add value to all aspects of the business

FIGURE 9.2
Analytics project needs.

If we look at the analytics projects' data requirements, most of the information that they need should be already available with the data warehouse team. The project team does not need to reinvent the wheel; they just align with the data warehouse group. Additionally, the data warehouse team has tremendous capability in moving data around, applying data cleansing techniques, aggregating and summarizing data, managing histories, etc., and have robust and scalable infrastructure in terms of CPU and storage. They may also have data privacy and security controls in place across their data supply chain. Analytics needs all of that; Figure 9.2 is a snapshot of the overlap between data warehouse capabilities (see Figure 9.1) and the capability needed for an analytics program.

METHODOLOGY

The analytics implementation methodology, therefore, relies on the data warehouse infrastructure, processes, and technology, and introduces the advanced layer of tools and human skills in analytics modeling and decision strategy as

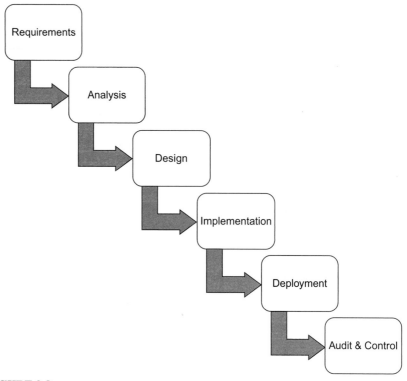

FIGURE 9.3
Analytics project life cycle.

per the Information Continuum. The methodology presented in Figure 9.3 follows the traditional waterfall approach, but then it contrasts tremendously with traditional software development. Individual situations may demand adjusting this methodology for specific organizational needs or for specific projects.

Requirements

The entire Chapter 8 is devoted to requirements gathering, emphasizing the importance of requirements in an analytics project. The four parts of the requirements gathering or extraction process result in a requirements set that includes the following:

- Problem statement and goal
- Data requirements
- Model and decision strategy requirements
- Operational integration requirements

These categories make up the requirements section of the methodology, and since the detail has been covered in Chapter 8, we will not further elaborate here.

Analysis

The purpose of analysis is to describe the requirements in greater detail with the context of existing business processes and data utilization, and include all the moving parts and components needed for delivering on the requirements. When gaps are identified and clarifications are needed, the business community is engaged and clarification is sought. The system, process, and requirements analysis allows for a holistic understanding of the overall problem and its potential solution, so the boundaries for design and development can be set. The analysis is also responsible for identifying the easier and challenging parts of the project, so appropriate staffing or consultative help can be sought and the feedback on general timelines can also be established once the analysis is concluded.

The following categories of analysis need further elaboration:

- Problem statement and goal—analysis
- Profiling and data—analysis
- Model and decision strategy—analysis
- Operational integration—analysis
- Audit and control—analysis

Problem Statement and Goal—Analysis

This is the most important part of the analysis stage of the methodology, as to if what the business is trying to achieve is even possible knowing the available technology and skills. The analyst should have a solid handle on the various analytics techniques that will be used to achieve the stated goal. A review with the technology of that goal is also critical because sufficient historical data may not be available, the analytics method to be used may not be present in the technology stack or skills, and resources may have a gap. The analytics analyst, therefore, has to validate, clarify, and confirm the following.

Can the problem statement and goal in fact be solved by the available analytics methods? Are the methods available in the technology suite that the analytics team has and is there enough experience available within the team to undertake the analytics project? At times, incorrect expectations or misinformation result in broad assumptions about the capabilities of the analytics technology.

Profiling and Data—Analysis

Data requirements are also covered in detail in Chapter 8. Through business processes, the data that gets generated may already be available in the data warehouse. In the data analysis stage, data profiling of that data will be carried out both in the operational (source) system and in the data warehouse system. There are two categories of data profiling activity: syntactic and semantic.

Syntactic Profiling

This type of profiling deals with one field at a time and profiles are established on each field. While it can be carried out on all fields in the scope, that may be overwhelming. Therefore, this profiling should be limited to data fields of interest. The data fields of interest are those that the business routinely deals with in reports, KPIs, and other performance metrics. Some ideas on spotting these can be low cardinality fields like types, statuses, and codes, or cities, countries, and currencies. Low cardinality means that the total number of possible values in that field is relatively small compared with its occurrences in records. For example, we may have 100,000 customer records but the field named "Gender" may only have three to four possible values (Male, Female, Unknown, and Null), therefore, there are many repeats for the 100,000 records. High cardinality values are usually bad for profiling unless they are numeric and can be aggregated to represent averages, minimum and maximum like volumes, amounts, counts, etc. The sales amount per sales transaction is a good field for profiling, but a VIN number is bad along with various types of account numbers and IDs.

It is important to understand that profiling is not being done for data-quality or analytical reasons. It is being done to develop better understanding of the data. The quality component should already be covered within the data warehouse program when the data was acquired. This understanding will come in handy when candidate variables are being identified for analytics model development and used in decision strategy designs. The information gathered as part of syntactic data profiling is presented in the following list; not every profiling attribute is collected for every type of field. The profiling can also be done on a sample of the total records and not necessarily on multibillion row tables. Most database systems, reporting and analytical systems, and data mining systems provide some kind of automated capability of data profiling. Advanced profiling tools can also be acquired for this purpose or custom built for Big Data challenges.

1. *Minimum.* The minimum possible value in the data set for that field. This is not applicable to character type (string) fields but numbers and dates should have this.
2. *Maximum.* The maximum possible value in the data set for that field. This is not applicable to character type (string) fields but numbers and dates should have this.
3. *Mean.* Similar to minimum and maximum, the mean is the average and will only apply to numeric values.
4. *Median.* The value that lies in the middle of a list of possible values. The values have to be sorted and counted, and then the median is value at the center of that list.

5. *Mode*. The highest repeating value in a list of possible values.
6. *Standard deviation*. A good measure of how aligned the overall values are to the mean. Higher standard deviation means the values are spread out and the field has high dispersion, while small standard deviation means that the values are closer to the mean.
7. *Frequency*. Frequency is the count of number of occurrences of the same value. So for a field called Ticket Status, if there are a 100 records and 40 are booked, while 55 have not been purchased yet and 5 have been purchased and cancelled, the frequency of the field Ticket Status would look like the following:

Value	Frequency
Booked	40
Not Sold	55
Cancelled	5
Unknown	0

The value for Booked has the highest frequency. For high cardinality fields, the top five values are sufficient for frequency calculation. In this example, we had 100 records and all of them are accounted for since the frequency sums up to 100. If we were looking for the top two frequencies, we would get Booked and Not Sold and that would've represented 95% of the data.

Frequency is not limited to any particular data type. Also included are:

1. *Distinct values count*. The count of distinct (unique) values stored in a field. In the preceding example, the distinct values count for the field Ticket Status is 4. This is equally applicable to numeric, date, and string fields.
2. *Maximum length*. This only applies to character (string) fields and it is the size of the values in the field, not necessarily of the field itself. A field may have a data type of character (20), but upon reviewing the data it turns out that no value is over 10 characters long. The minimum length is typically not a useful profiling attribute.
3. *Null*. This covers how many values are actually null, meaning the value is not available. Null counts are also applicable to all data types.
4. *Zero*. Zeros apply to numeric fields only and counts how many times zero appears as a value.
5. *Time series*. This is a yes/no profiling attribute and indicates if a field has some time-series value. Sales Amount is a meaningful time-series field since as it can be trended over time. Currency Code is not a meaningful time-series field.

The preceding information should be recorded in a formal document that will be used in design input. In addition to profiling attributes for each field, a detailed definition should also be captured that should cover:

- The business definition of the field.
- How the field is created as part of the business process.
- How the field is modified/deleted as part of the business process.
- How the field is used in the Information Continuum.

Semantic Profiling

The semantic profiling of fields tries to establish their interdependencies and correlations in a business context. For any data set that has some kind of hierarchy, such as Customer → Account → Order → Transactions, the customer has accounts and those accounts are used to place orders that are then paid for by financial transactions. This type of hierarchy with their counts in the source or data warehouse system should be tracked. For example, there are a total of 1 million customers in the source system; a total of 3 million accounts (meaning an average of three accounts per customer) have placed 12 million orders in the last three years with a total sales amount of 120 million. This provides interesting correlations within the data and a relationship is built within these data entities. Then, if one of them is forecasted through analytics, the others can be easily estimated.

The organization structure of a corporation is another example of hierarchy, where the corporate legal entity sits at the top, then business units and departments underneath, then functional teams, and eventually the workforce. In this type of hierarchy, just the relationships are important as to which teams are part of which business units, and not necessarily volumes. So there can be a volume-based hierarchy or a structure-based hierarchy; both establish dependencies between fields, which is part of semantic profiling. A brief about historical data availability is also a good profiling attribute that can come in handy particularly when building forecasting models.

Model and Decision Strategy—Analysis

During the problem statement and goal analysis, we covered the analysis of the analytics method and its applicability to the problem. The model and decision strategy requirements are analyzed in this stage in light of the data analysis covered in the previous section. Every single field candidate to be used in the model as a general thought process has to be reviewed against its data profile. Fields that have skewed values, such as 90% of the field Gender, actually contain the same value (Male) and therefore it is not a good candidate for model development. Although it may be a field extensively used in decision making within the business community today, now the analysis reveals that it cannot be used in the analytics model.

Similarly, the decision strategies may be relying on decision variables (see Chapter 5) to break down the decisions and spread the workload across a larger workforce, and if the decision variable (the field on which segmentation is performed) is skewed, then most of the workload may go to the same queue. Let's use an example of a loan collection system. A predictive model can assign a probability of "collect-ability" to all accounts in collection. Business decides that everybody over 70% collect-ability will be assigned to the internal collection staff, while the lower collect-ability accounts will be handed over to a third-party collections agency. If the collection department had 20 analysts, the cases with over 70% collect-ability now need to be further segmented on additional variables. Usually this input comes from the business, but after the data profiling, that has to be analyzed again to see how the decision variables are spreading the data.

Nulls and zeros are also important, as they have limited use in model and strategy decisions and have to be taken back to the business to explain and seek alternatives.

Operational Integration—Analysis

The analysis around operational integration is very similar to a traditional system integration analysis where systems, their interfaces, their technologies, and their timing are analyzed, understood, and documented. Typical requests from business are always for real-time availability of information in other systems, but the real-time integration has some cost associated with it and, therefore, the business has to be probed on why real-time integration is necessary. In the case of the preceding collections example, a nightly integration from the analytics system into the collections workflow and queuing system may be alright since workers come in the next morning and their work queues are preloaded.

For decision strategy integration into the operational system, the following key points have to be analyzed and reviewed with IT and business:

- What is the event that would trigger the analytics module and the decision strategy to come into action? A new insurance policy request, a new phone connection, a missed payment, an insurance claim submission, a new order booked, etc. are all business process events that can trigger an analytics module and a decision strategy.
- What is the data that will be passed to the analytics module? How complete is the data going to be at that time? It may be that waiting a day on the data allows for more complete information to be available in the system, for example, a field staff enters data manually toward the end of their trip.
- Is a final decision being passed back after the strategy or just the analytics output? In case of FICO scores for consumer lending, typically a score

is received from the credit bureau (e.g., Experian) and then the decision strategy is built into the operational system on that score.

- Decisions received or computed by the operational system have to be stored somewhere; the analysis will highlight this aspect of the analytics solution. The storage of the results can be within the operational system or outside it, but the analysis has to document the details for designers.

Audit and Control—Analysis

This brings us to the last aspect of analysis—audit ability and control. The business would require some ability to audit the historical decisions recommended by the analytics solution. How exactly will that data be recorded, where would it be kept, and how would the auditors get to it? Does it fulfill internal control requirements or industry-specific requirements as well as that of regulators? Who exactly updates the audit information and how is it made temper-proof? Since the analytics-based decisions may need to be reviewed from an efficiency perspective (how effective is the analytics module) as well as from an audit perspective, should this data be stored in one place or two places? This is the line of questioning the analyst has to follow. It may be tough to get the answers to these questions depending on how mature an organization is toward analytics for their day-to-day decisions. The analyst, therefore, will have to educate and build consensus across departments about how to audit the automated decisions taken on the output of analytics models.

Design

Similar to the analysis phase, all the various categories that were analyzed now need to be designed. The design of an analytics solution has six components:

1. Data warehouse extension
2. Analytics variables
3. Analytics datamart
4. Decision strategy
5. Operational integration
6. Analytics governance—audit and control

Data Warehouse Extension

Based on the data requirements and the analysis of the data fields, one of the gaps is data availability in the data warehouse that needs to be designed. The design principles for this additional set of data from various source systems should be done exactly according to the process and methodology in place for data warehousing. They bring in thousands of fields from various source systems and have the technology and skill to do that for additional data of interest for analytics. It will add to their existing data structure as more data is added, and the analytics project team will not have to go through the learning curve of accessing source systems, staging the extract data, nightly scheduling of batch jobs, ETL tools and technology, and the database infrastructure. The data warehouse extension design will have a data modeling piece and an ETL piece that will be responsible for bringing the data in and keeping it up-to-date on a scheduled basis as part of the overall data warehouse maintenance process.

Analytics Variables

The variables are where the art and science come together. What creative variables can you invent that may have a high analytic value? There is a difference between variables and fields. In data warehouse systems, when an analyst goes to a source system to look for data that is needed by the business for reporting and analysis, two best-practice rules are followed:

- The first rule deals with trying to get the most detailed level or atomic data possible. If the business has asked for daily sales data per customer per store, the source system usually has the line-item details of what the customer bought and how much it was. The most granular data would be at the line-item level in the basket of the customer, and that is recommended to be sourced and pulled into the data warehouse.
- The second rule deals with a principle called triage when sourcing data. The data that is needed by the data warehouse driven from business requirements is priority 1. Then there are operationally important fields like CHNG_UID, which is a field that captures the user ID of the user who last changed a record; these are priority 3 fields. Everything in between is priority 2. The best practice is to pick up everything with priority 1 and 2 while you are in there building an extraction process. It may come in handy later.

These rules are why the data warehouse is supposed to have more fields than seem necessary for the analytics use. Going back later and getting the additional fields is far more difficult than keeping the additional data around just in case. The analytics project can actually get lost in this large river of fields and not know which ones would be valuable. Following are the four kinds of variables to help sort through the list of fields, and also an explanation on how to use these variables because their treatment changes through the project life cycle.

Base Variables

Base variables are the important fields present in the data warehouse within the analytics project's scope. If the project's scope is to build a sales forecasting model, then the business context is sales, and therefore all the fields in the data warehouse within or linked to sales are potentially base variables. If the sales department's users access certain dimensions and facts for their normal business activities, then everything in those dimensions and facts is potentially a base variable. Examples are store types, customer age, product weight, price, cash register type, employee code, etc.

Performance Variables

Performance variables are specialized variables created for analytics models. If the project's problem statement is a predictive model that calculates the probability that a customer will use a coupon to buy something (propensity modeling) or the probability that an order will ship on time, then it would need the base variables (formatted and transformed) as well as some new and interesting innovative variables that may have a high predictability for the problem statement. The base variables are raw data and usually have continuous values like age of a customer. The performance variables are preferred to be coded variables. So if the customer records have age as follows, 28, 33, 21, 67, 76, 45, 55, 68, 23, etc., then the coded values would replace the age variables as Code 1 (referring to ages less than 21), Code 2 (referring to age ranges from 21 to 38), and so on, and the last one would be Code n (age greater than 100). This way the age distribution, frequency, and its role in the predictive model can be analyzed and tuned.

Other performance variables could be as follows:

- Total sales of grocery items
- Total number of year-to-date transactions
- Percentage when credit card is used for payments
- Online user (yes/no)
- At least one purchase of >$100 (yes/no)

These performance variables are designed looking at the problem statement. There is no well-defined method to what should be a performance variable. In established industries like customer analytics (within marketing) and consumer credit (risk management), the analytics experts know what performance variables are going to be important, but in other industries like education, shipping and logistics, state and local governments, etc., the performance variables have to be worked out through trial and error over time. Some may become useful and some may not have any predictive value. Chapter 11 deals with this in greater detail. However, they have to be designed so they can be implemented and therefore they are covered in the design stage.

Model Characteristics

The variables (base plus performance) that end up being used in the actual model get a promotion and they are labeled characteristics, as they will get weights, scores, and probabilities assigned to them in the model. While the design may not know exactly which variables are going to become characteristics, there should be a provision to have some characteristics stored as input and then the model's output also stored with the characteristics. This is important for tuning and tracking the results of the model. The design has to accommodate characteristics (their creation from transactional data and the analytic output predicted, forecasted, optimized, or clustered).

Decision Variables

The decision variables were covered in detail in Chapter 5. These variables are used in decision strategies once the model output has been assigned to a record. So in case of manufacturing and prediction of warranty claims, if the model output assigns a 63% chance of a warranty claim against a product just being shipped, there may be 10,000 products with 63% or higher, so not all can be kept from shipping and not all can be inspected. Therefore, additional filtering is needed to separate the 10,000 products into smaller more meaningful chunks against which actionable strategies can be carried out. For example, if the product is part of a larger shipment, then let it go as long as the prediction is greater than 63% but *not* greater than 83%. This additional filtering is called segmentation in a decision strategy and the variables that are used to apply this are called decision variables. The importance of decision variables is establishing their thresholds for actions. In this example we have stated that if the product is part of a larger shipment, what does "larger" really mean—100, 500, or 1,200? These thresholds cannot be set arbitrarily and there is some structure to setting these segmentation values.

Analytics Datamart

The analytics datamart is a specialized database designed to handle all types of variables, their histories, their revisions (in business rules and calculations), and their structural dependencies (parent–child). The input into the model, test data, training data, output and validation results, and the model generation are all maintained in this datamart. The label "datamart" should not be confused with a dimensional star schema type mart (Kimball, 2002); datamart is used to refer to a specialized subject-area focused collection of relevant data. Since it hangs off the data warehouse, it has been labeled as a datamart. Its structure is very different from that of a traditional datamart and more innovative and out-of-the-box type designs may be necessary to meet its functional purpose. An experienced data modeler should be able to fully understand the purpose of the data in this collection and model to meet the requirements.

Decision Strategy

The decision strategy design is covered in detail in Chapter 5, therefore only a brief guide into its design will be covered here. If the requirements layout of decision strategy is detailed enough and all metrics, decision variables, their thresholds, filtering conditions, and actions against strategies are clearly articulated, then there is really not much to design. If a strategy design tool is available, then even the implementation part is simplified. The design step has to ensure that the business rules are defined quantitatively, so statements like "if the price is too high…" are in requirements, the design step will force the business to pick a number as to how much is "too high." If the business doesn't know, then the data profiling carried out during the data analysis will have to be shared so they can determine a quantitative equivalent of "too high" (e.g., greater than 4,000).

Operational Integration

The design of the operational integration has three pieces:

1. Strategy firing event
2. Input data
3. Output decision

Strategy Firing Event

This is the event that will cause the strategy to be fired, meaning executed on a data set. There are two types of strategy events:

1. *Business process event.* These are real-world business process events that can fire a strategy. An example is a loan application. An online loan application is submitted from a bank's website and the data is passed to the analytics model for default prediction. Based on the probability of default, the strategy is fired to see what kind of pricing, fees, and other terms can be offered if the case is approved at all. There is no controlling when such an event occurs and strategy execution is therefore in real time, driven from the loan application event generated by the customer who visited the bank's website.
2. *Scheduled events.* These are batch schedules where large amount of transactions are passed through the analytics model and the decision strategy and all of them are assigned an action and activity (the decision). An example of this would be a business scenario where transactions come in throughout the day but the processing starts with work assignments the next day on that data. Take college applications where the college staff wants to focus on applications with a higher probability of acceptance. Applications pour in the last day of the deadline and a nightly schedule assigns all of them with a probability; the decision strategy breaks the applications into buckets, which are reviewed the next morning by the staff. The same goes for insurance claims where a fraud probability is assigned to incoming cases and senior claims processors deal with higher fraud probability claims the next morning after a decision strategy decides how to assign the case workload.

Both of these scenarios need to be understood and the integration is designed accordingly with the operational system.

Input Data

The input data is the definition, format, and layout of the data that is passed to the analytics solution (model and decision strategy) by the operational system where the event-based or schedule-based transactional data resides. Careful attention has to be paid while mapping the variables in the model and decision strategy back to the data in the operational system. There will certainly be need for transformation of the data coming from the operational system, and it is up to the designer to either perform the data preparation (from fields to input variables and characteristics) in the analytics solution or in the operational system before data is sent. The data warehouse processes should be tightly aligned and leveraged as much as possible instead of reinventing the wheel.

Output Decision

An output decision is an actionable code that translates into some activity in the operational system. What to do with the output has to be coded in the operational system and, therefore, operational system engineers should be part of this design discussion. When the analytics solution sends an "approved" code back for an insurance claim or a loan application, what should be done within the operational system on that information is a design question requiring close business process input as well as operational system changes. Without this design piece, the decisions may get assigned but never actually carried out.

Analytics Governance—Audit and Control

The audit and control design regards what information is to be stored related to what data was sent in, what probabilities or other analytics output (cluster, forecast values, etc.) were assigned, what decisions were determined to be applicable, and how the decisions were carried out. This is not a simple design, and audit framework has to be understood before attempting to build a transparent analytics solution. Chapter 6 covers this in detail and, therefore, it will be briefly covered here. For the most part, once the model and strategy is in production, there is limited manual intervention and manual oversight of the transactions passing through the solution. An audit is therefore the only function where things can be tracked, monitored, and thresholds set up so automated decisions can be stopped or investigated. This same type of information is also needed for strategy and model tuning, and therefore the design for audit and control should accommodate both sets of requirements as far as storing the information is concerned, even though its use may be different for different sets of users. Even if the data is kept separately, its structure and processing should be done consistently within the same design.

Implementation

Implementation deals with the software program's development that will be processing data. The bulk of the implementation is actually ETL, meaning taking data from one place, manipulating it, and then loading it in another place. ETL programs are data-centric and do not involve development of any workflows or screens and user interactions. They are usually SQL and other data cleansing, formatting, and aggregation programs written in a wide variety of supported languages that are implemented and operated like a blackbox. ETL implementation includes sourcing data from source systems or from data warehouses all the way through to the analytics datamart, as well as the implementation of audit and control data recording, pulling from the integration inputs, outputs, and decision strategy segmentation.

The implementation of an analytics datamart should be like any other datamarts within the data warehouse, and no specialized tools or technologies should be needed. Therefore, the first two pieces of implementation are the analytics model and the decision strategy; the third one deals with governance. The analytics model is implemented in a data mining tool or in data mining libraries available within the database engine. Specific knowledge of that tool is essential for the implementation and specialized training is required. Ordinary software development resources will find these analytics programs very difficult to understand and manage. The terminology, concepts, and the model development steps, its testing, validation, and performance evaluation are all very specialized areas, and people are required to have formal training in data mining implementations through undergraduate or graduate classes in advance programming and artificial intelligence courses. Advanced knowledge of statistics can also be very useful but not required.

The implementation of a decision strategy has several options. It can be implemented in traditional programming environments like Java and DotNet; it can be implemented using specialized strategy management tools available from vendors like FICO and Experian; or with business process management (BPM) tools. Even ETL tools are capable of implementing decision strategies at a basic level. The second implementation piece is the integration of the analytics solution and the operational system, and it will be carried out using any supported environments already in place, such as a messaging or service bus integrating the operational systems. The last implementation piece is that of audit and control data access. The audit and control data is loaded using ETL, but then some reporting and analysis is required to set up thresholds. This can be done either in a database development environment or with specialized components available in OLAP tools already in place with the data warehouse, such as a dashboard.

Deployment

As part of the implementation, the analytics model is validated and business starts to believe in the model's ability to help make business decisions. At this time the model is ready for deployment, so transactions can be passed to the model and strategies can be executed on the model output. This integration is part of the deployment phase. The model implementation is tackled separately from the decision strategy implementation because of the difference in technology and toolsets used for each. No data mining tool or library has built-in capabilities for running a strategy because of the difference in their technological nature.

Once the two are integrated, the combined effect has to be tested and validated on actual data. Since a strategy-based business process activity is usually a new paradigm for the business, it takes a while and a lot of trial and error before strategies perform at a level where business sees value in their output. This step can be quite complicated for larger user base deployments—that is, where a larger number of users are awaiting their work queues to be adjusted based on analytics model output. Areas like price optimization may have an easier time since not many people need to run various pricing models and evaluate output. Once the integrated deployment of the analytics model and strategy is tested, validated, and approved, the operational integration is hooked in for complete end-to-end integration testing of the analytics solution, including the audit and control data recording and review.

Execution and Monitoring

The last step of the analytics implementation methodology is execution and monitoring. Once the solution goes live and automated decisions are carried out using the analytics model and strategies, close monitoring is essential to see when the model may need tuning or decisions thresholds adjusted. Also important is the champion–challenger notion of new decision strategies where an existing strategy is champion and a new strategy is tested to see if it improves results. The new strategy becomes the challenger. If after some observation it appears that the challenger strategy is performing better, then it replaces the champion strategy.

This is an ongoing process and the monitoring of expiring strategies and their performance data has to be retained to review the basis on which it was expired. As organizations get better at building challenger strategies, more innovative and specialized improvements are introduced into the decisions and business processes become smarter, reacting to specific customer behavioral changes, economic and market shifts, industry and regulatory changes, as well as internal cost and efficiency drivers.

CHAPTER 10

Analytics Organization and Architecture

This chapter covers two topics—organization structure and technical architecture—the final pieces in building a strong team and foundation for analytics projects in an organization. The purpose and theme of these two topics remains consistent with the rest of the book and are focused toward simplification and democratization. Both the organization structure and technical architecture are built on top of the existing data warehouse organization and architecture.

ORGANIZATIONAL STRUCTURE

The methodology presented in Chapter 9 covers the steps needed to implement analytics projects. The idea is to have a hybrid (centralized/decentralized) team structure that can deliver analytics projects for all departments as a continuous activity, even handling multiple projects at a time. An organization structure required to accomplish that would be a matrix organization and is built combined with the data warehouse team structure. This would set up a business intelligence competency center (BICC) responsible for implementation, including data warehousing, analytics, and decision strategies across the entire Information Continuum. It is highly recommended that analytics teams work in close proximity to the data warehousing teams because of the overlap in knowledge, skills, and technology. If the enterprise data warehouse team already has a formal structure closer to the layout of a BICC, then it should be simply extended to include analytics implementation as part of their responsibility.

The following are some suggestions and explanations for organizations to adjust their existing environments. It is not required to whole-heartedly adopt this exact structure, but it is important to understand the dependencies and overlap, and then work out a plan to slowly converge toward this end-state.

CONTENTS

BICC Organization Chart

Here is a proposed organization structure of the BICC (Figure 10.1) with five centers of excellence dealing with:

- ETL (the overarching term referring to all aspects of data in motion)
- Data architecture
- Business analysis (collectively refers to data, analytics, and requirements analysis)
- Analytics
- Information delivery

This is a direct reporting hierarchy with BICC having management responsibilities and budget for this team, while a charge-back structure could be put in place for project assignments. Analysts prefer to have direct reporting into the BICC, but they should have a dotted line back into the business functions as a liaison to develop specialization in the functional areas and understand the opportunities and difficulties. This will help ensure the application of the Information Continuum can be identified at the appropriate level and a project created and delivered for ongoing value. This would also allow the analysits to become more familiarized with the technology and implementation complexities.

In addition to the core team of the BICC, there are additional skills needed to effectively build and deliver analytics solutions (Figure 10.2). These skills are

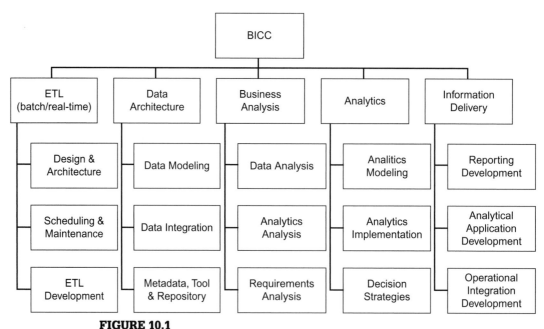

FIGURE 10.1
BICC organization structure.

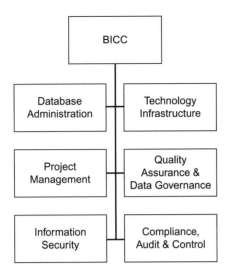

FIGURE 10.2
Additional teams.

enterprise in nature, meaning other aspects of IT development and support also share these skills. Since analytics projects will always put IT infrastructure to the test with large historical data sets, complex transformational programs, and massive data processing throughput needed to build and test models, these additional skills need close alignment with the BICC. These include:

- Database administrators (DBAs) responsible for database management.
- Technology infrastructure (dealing with data centers, servers, storage, etc.).
- Project management. Analytics projects are going to need project managers and, instead of building their own capability, some designated project managers specializing in analytics should be used who are part of the central project management office.
- Quality assurance (QA) and data governance. Similar to project management, the centralized QA team responsible for all development activities across IT should be used instead of building one from scratch with some specialization analytics.
- Information security. It is expected that a robust data warehouse implementation will have addressed the information security framework around data and reporting. That capability should be extended into analytics and decision strategies.
- Compliance, audit, and control. Both compliance and internal audit teams have some oversight on data warehouse although their focus is typically operational systems. However, unlike data warehouse, an analytics project involves decision strategies that impact day-to-day business activities, and therefore a new specialization within compliance and internal audit teams may be required.

The BICC should be connected to these additional organizational capabilities through projects, where each analytics project engages these teams and ensures the project is aligned with their mandate. In situations where that is not possible, such as technology infrastructure and DBAs who have to engage at all times to keep the deployed systems running, a formal roles responsibility and handover mechanism should be developed.

Roles and Responsibilities

The material in this book does not necessarily concern itself with data warehousing teams, skills, and implementation. Rather it recommends that existing building blocks already in place within data warehousing teams, technologies, and databases should be used to fast-track implementation of analytics projects and therefore build on top of a data warehouse. With that approach it would make sense to have the new analytics organization tightly knit into the data warehouse organization. However, the roles, responsibilities, and deliverables that are specific to data warehousing are not ironed out in detail. A lot of good material is available from authors like Ralph Kimball (2008) and Bill Inmon (2005) to cover the detail on the data warehousing side. This section will only cover the roles and responsibilities as they pertain to analytics projects.

ETL

As mentioned in previous chapters, ETL stands for extract, transform, and load, and it is used as a noun in this book. It is a term that was coined in the mid-1990s with the advent of data warehousing. The idea was to take operational data out of the transaction systems and move it into a separate database called a data warehouse for reporting and analysis. This massive undertaking of extracting all the data out of the operational systems, most of which were legacy database systems that relied heavily on file-based structures, into a new data warehouse required a methodical approach and a tool that simplified the task. Therefore, *extract* (from files or legacy database systems), *transform* (into a more integrated and structured form using relational databases and applying data quality and business rules), and *load* the data into a data warehouse.

In the context of this material and the modern shifting definitions, the more batch and file–oriented ETL has been transformed into a more robust high-performance, parallel-executing, real-time integration suite. So the term *ETL* is used here to refer to the entire capability of moving data of all sizes in a fast and reliable manner from one place to another—that is, data in motion. The data could be in batch or real time, scheduled or event driven, files or single transactions, and also provide audit and monitoring capabilities. This can be achieved through an integrated suite of tools from one vendor or a collection of tools from multiple vendors.

Within ETL, there is always an architect who is responsible for designing the overall environment and looks at all the jobs, their dependencies, their error handling, their metadata, etc. ETL has two flavors: the design and development, and the scheduling and execution. The architect is responsible for designing the environment so consistent development methods like shared staging area, common methods of key generation, look-up of shared referenced data, retention of processed data, or files and naming conventions are consistently followed across all types of data movement activities going on within various projects. The other piece of an architect's role is performance, job dependency map, and the error and restarts design necessary for a large and complex ETL schedule. The flavor of ETL specializing in analytics solutions has been covered in Chapter 5 on decision automation. The ETL team is an essential part of any data warehousing operation, and they usually have a preferred ETL tool to design, develop, and execute ETL routines to move and manipulate data. The trick is to break down the specific data movement tasks within an analytics solution and hand the tasks to existing teams that specialize in certain areas. However, one ETL team should deal with all projects across the entire Information Continuum.

Data Architecture

The primary responsibility of the data architecture team is building data models. Ideally, the same team should be building data models for all types of systems across the entire Information Continuum, from operational reporting to data warehousing and analytics. A data model is at the center of data movement and data access, and therefore it is best positioned to define and manage the overall metadata for any data-centric project. Some examples of metadata that needs to be tracked are:

- Source system field definitions.
- Source system business process to field mapping.
- Source to target mapping.
- Any transformation and business rules at the field level.
- Aggregations and performance variable definitions.
- Mapping of fields from the source system, to the data warehouse and reporting, to the analytics datamart, to the predictive models and decision strategies, to auditing and monitoring the database (called data lineage).

Since the data model inherently captures the data dependencies and brings together the nuances of various forms and data types, the data integration rules have to come from the data model. Without fully understanding the data, models cannot be built, and without a target model, transformations and integration logic cannot be built. Therefore, it is recommended that these three functions belong in the same team. More detailed data modeling techniques and theory around abstraction and database designs are beyond the scope of this book. If a robust data warehouse is in place, then a mature data architecture team, technology, and process have to be in place. Investment in metadata management is always a tough business case, and therefore is practiced in a fractured and sporadic fashion in most data warehouse implementations. In analytics projects it is central to testing and validation of decision strategies where the entire Information Continuum comes together to carry out a business decision with little to no human intervention. If the knowledge of the data fields, their origin, and their transformations are not evident, the business will never sign-off on trusting the automated decisions.

Business Analysts

Business analysis and requirements gathering or extraction for analytics projects have been covered in great detail in Chapter 8. Traditional business analysts within data warehouse teams, their skills, and best practices are beyond the scope of this book; that material is extensively covered by authors like Bill Inmon (1992) and Ralph Kimball (2002) among various others.

Analytics

The analytics group is a new addition to the data warehouse team and capability. Within this group the following new capabilities need to be established:

- Analytics modeling
- Analytics implementation
- Decision strategies implementation

Analytics Modeling

The analytics modelers have traditionally been statisticians usually with advanced degrees in statistics and mathematics. The financial sector has attracted a huge chunk of this knowledge and skills base in the last 15 years. A wide range of applications in forecasting, prediction, and decision optimization have been in place since the 1950s (May, 2009) where these modelers (also known as quants within the investment banking community) have been building statistical models for economics, finance, manufacturing, etc. They are very expensive and are usually highly specialized in a particular problem domain. In large organizations they appear as a silo with a magic box from where decisions come out. Financial losses, rogue traders, and systemic risks are issues that crop up when these modelers work with a black-box approach with limited visibility, understanding, or transparency for management. To understand and audit what they do, you have to have an advanced degree in statistics.

The advent of artificial intelligence has created an alternate school of thought where machine learning and software based on data mining techniques are able to tackle newer problems having unprecedented data volume and processing requirements. Data mining underneath is still mathematics and statistics implemented through software programming, but it abstracts the underlying mathematical complexity from the user. The current statistics- and mathematics-based analytics include disciplines such as:

- Linear/nonlinear programming and numerical analysis or even differential equations
- Other areas of applied mathematics like game theory and cryptography
- Constructs from Euclidean geometry
- Stochastic problems
- Multivariable regression
- Probability theory

The use of analytics modelers has been limited to a handful of specialized areas and they are never embedded in a data warehouse team. They tend to do their own data sourcing, their own evaluation of variables, build their own analytics datamarts as standalone systems, and build and deploy their own models. The decision strategies are typically communicated as a document or as a set of instructions to the staff responsible for making decisions. In this climate, they cannot tackle Big Data problems because of the sheer size and scale and also work with well-defined variables within their problem domain. It is extremely difficult for them to try out new data sets and new variables just to see if something useful jumps out, as their skills are confined to a handful of people who complete the end-to-end tasks.

The comparison in Figure 10.3 shows how the current analytics environment functions and why it cannot scale to either the Big Data challenges or democratization demands from other business units. We have been arguing and making a case for the alternate approach as it is easier for adoption across all parts of the organization. The statistics-based specialized analytics modeling will remain confined to a handful of areas where organizations have the appetite for having a dedicated solutions team. That would be outside this organization structure. The analytics modelers are the people building the actual models using data mining software, and therefore need expertise in the tools being used in the organization.

Analytics Technology

The analytics toolset along with its specialized databases, audit and control requirements, and the metrics and thresholds require a full-time analytics team to manage the technology. This team will work closely with the data warehouse technology teams and the infrastructure teams. If specialized analytics software has been purchased like SAS (2012) or SPSS (2012), this team will be responsible for installation, support, and ongoing production

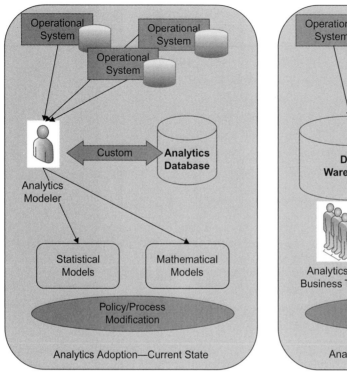
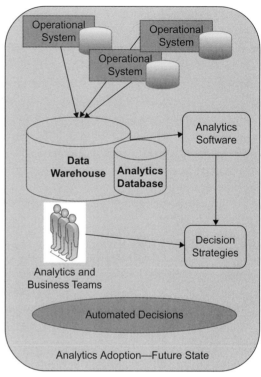

FIGURE 10.3

Comparison of analytics implementation process.

monitoring and operational SLAs for that software, as well as development support, tuning, configuration, and deployment of models.

Decision Strategy

A specialized team will be needed to support the existing and new strategy development, tuning, and audit. If a decision strategy tool is purchased, the analytics technology team will support the tool while the decision strategy team will be its user community. This team will work closely with the business and help them build strategies and provide feedback on strategy performance. They will maintain the generations of strategies and keep running champion–challenger analyses for the business, as well as carrying out simulations on decision strategies.

Information Delivery

The information delivery group or team is usually in place within the data warehousing team and is responsible for report development and execution.

Two new additions to this team are necessary: one for specialized analytical application development like visualization and GIS mapping–based analyses, and the other for developing the integration with the operational systems for strategies and analytics output. Report and analytical application development is out of scope of this book, however the operational integration of decision strategies has been covered in detail in Chapter 5 on automated decisions.

Skills Summary

The new skills required above and beyond data warehousing skills would be in the following areas.

Analytics Analyst

The analytics analyst will have to have a deep business understanding before a conversation can be initiated with the business. The following are skills needed by an analytics analyst:

- Business domain knowledge
- Analytics problem patterns (as suggested in Chapter 3 on using analytics)
- Understanding of the four analytics methods
- High-level business process modeling
- High-level data profiling ability

Analytics Architect

The analytics architect's role is an extension of the data warehouse architect role. The conception of the overall analytics solutions, including data from the data warehouse, design of the analytics datamart, implementation of decision strategies, and operational interfaces, all need to be holistically placed in one solution. Additionally, there may be multiple solutions being designed and implemented simultaneously, so it is important to ensure that resources are not overlapping and duplication of effort around same data is not occurring.

Analytics Specialist

The analytics specialist is the analytics modeler and should be well versed in the toolset being used, be it a specialized software or database in-built analytics. Additionally, the analytics specialist should have extensive database knowledge and the ability to guide developers to write programs for data cleansing, data manipulation, sampling, data integration and aggregations, etc. On top of that, the most important skill is the ability to validate and test models, identify their weaknesses, add and play with additional variables, and tune parameters available within the analytics software to build a robust model.

TECHNICAL COMPONENTS IN ANALYTICS SOLUTIONS

Figure 10.4 depicts a typical technical architecture for an analytics solution built on top of a data warehouse. The implementation detail of all the components in this figure will not be discussed in detail as data warehousing resources are filled with in-depth design and implementation guides for these components. We will only concern ourselves with the analytics datamart.

Analytics Datamart

The term *datamart* became part of the mainstream data warehousing and reporting industry right after Ralph Kimball introduced the world to dimensional modeling (Kimball, 2002). It is defined as a subject area–specific collection of dimensions and facts like a sales order datamart, a billing datamart, an employee benefits datamart, a loan datamart, etc. It used to imply one fact table and various relevant dimensions. With design maturity in the industry, datamarts could handle multiple fact tables like aggregate tables in addition to fact tables at the lowest grain. An example of multiple fact tables within a datamart is a help desk ticket datamart where one fact table is at the ticket's

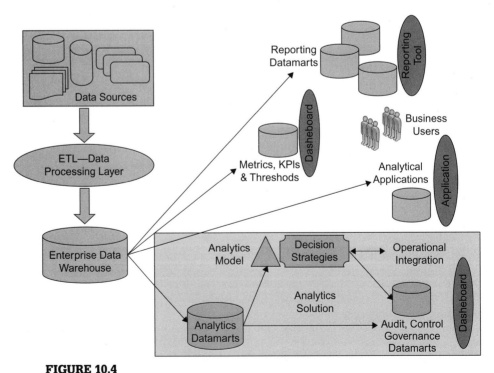

FIGURE 10.4

Analytics solution technical architecture.

grain (one record represents one ticket) and another can be at the ticket status grain. Datamarts built like that are read-only from a user perspective.

The definition of a datamart evolved with more specialized and data-centric software packages like anti-money laundering, pricing, or campaign management. There is usually a need for creating a separate database fully loaded with the relevant data needed for the software to operate. An ETL process is needed to extract data either from the data warehouse or from the source systems and load into this specialized database structured to support the software package. This is also known as a datamart because it is a specialized collection of data that can span multiple subject areas, but it is not necessarily built as a dimensional model and is used for more than just reporting. This definition of a datamart has data with both read–write functions and varying grains of data in the same datamart.

An analytics datamart refers to this later definition. It is basically a collection of all relevant data needed for an analytics solution to work. This would include the lowest grain detailed data, summary and snapshot data, performance variables, characteristics, model outputs, and the data needed for audit and control—any type of data relevant to the solution needs to be designed and stored in the analytics datamart.

ANALYTICS DATAMART SCOPE

An analytics datamart is a one-stop shop for all relevant data for the analytics solution. There are no specific design guidelines for building the analytics datamart.

The analytics datamart has four distinct subject areas within its logical construct. A solution-specific implementation may have a different physical manifestation of these four logical constructs. So a solution architect may decide to create two separate databases to house two logical constructs each. The constructs are:

1. Base analytics data
2. Performance variables
3. Model and characteristics
4. Model execution audit and control

Base Analytics Data

The base analytics data includes all relevant data needed for the analytics model. But what constitutes "relevant" is an interesting open-ended question and the first area of design that requires some art form in addition to the science. Let's say we are working on a logistics predictive model where thousands of shipment orders move through the system and we are trying to predict the likelihood of a shipment order being delayed. Every shipment order passing through the enterprise will be scored based on a probability of delay and higher probability orders will be handled through a dedicated service and handling staff to ensure appropriate communication to the client and possible rerouting. With this problem statement, what should be the relevant data that the analytics datamart should pull in from the data warehouse?

If we limit ourselves to order and routing data only, which would be the typical subject area definition for this problem statement, then we may or may not find good predictive patterns of data. It is recommended to use data from other subject areas to increase the breadth (number of fields) and depth (number of records) as much as possible to give the analytics algorithm a strong chance of finding the patterns most effective in predicting the outcome. In the regression-based traditional predictive modeling approach, a set of fields is evaluated for its discriminatory power in separating good and bad data (good in this case would be orders delivered on time and bad would be orders that were delayed). But exposing hundreds of fields to regression to find the good predictors and put their relevant weights into a model is a tedious iterative process, therefore usually that approach limits itself to well-known fields within the tightly linked subject areas. In this case, let's use the flow shown in Figure 10.5 to illustrate the life cycle of a package through the entire shipping process.

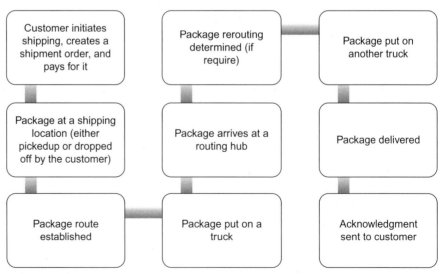

FIGURE 10.5

Flow of a package through a shipping process.

This simplified view of the package shipping process shows how the package moves through distinct stages from start to end. At each stage, a different set of data records are created and different data fields are populated or updated. The following subject areas (logical grouping of linked data) are in the shipping business process:

- Customer management (invoicing and payments)
- Package receiving (location, package label, invoice number, destination, etc.)
- Package routing (hops the package would take)
- Package transportation (truck, route, intermediate hops, tracking current location)
- Package delivery and acknowledgment

Throughout these data sets, there is overlap. For example, when a customer creates a shipping label at home and when the package is dropped at a location, the two data sets have some common overlap allowing for the company to link customer and payment to the actual package. Similarly, when the package is delivered, it requires a link back to the customer's email or phone to send the acknowledgment.

The definition of a subject area can be as granular as presented in Figure 10.5 or it can simply be just two subject areas: customer and package. Regardless of how many subject areas the data set is broken into, the combined data set is required for the predictive model to work. This is the science part of analytics. The art comes into play when you look at additional subject areas that may apparently have no direct impact on the delay of a shipment. This is what we would like the data mining tool to figure out. The additional subject areas could be:

- Human resources and shift assignments
- Truck and vehicle management
- Location or hub management
- Routing (or rerouting) parameters
- Contractors or outsource partner management

These subject areas are not actually tied to a specific package, but to the operation of shipping overall, and they can have an impact on the delay of a shipment. Sometimes looking for common patterns in delayed shipments is more difficult than finding the common pattern in on-time shipments. This is up to the data mining algorithm to figure out, and various implementations of data mining algorithms vary in their approach to this problem.

Therefore, base analytics data should include the preceding subject areas when data is pulled in from the data warehouse. If some of this data is not in the data warehouse, it should still be routed through the data warehouse using the common ETL, scheduling and governance teams, process, and infrastructure. Any time and resource considerations (coming from work-load on the data warehouse teams) should be trumped in favor of a more integrated approach, because creating a data silo sourcing directly from the source systems is going to have problems when we build performance vari-ables, build strategies, and try to prove that the model performs well. There can be additional subject areas available in the data warehouse, such as human resource data or the company's financial data, that one can argue may further help the model and should be included. This is a judgment call and no right answer exists. In this particular example it may be irrelevant and the other subject areas are enough to start the performance variable build process.

A few additional considerations for the base analytics data include:

- Data should be kept up to date and tied into the data warehouse load schedule.
- No new surrogate keys are needed, and the grain and key structure should be tied to the data warehouse.
- The design technique should also be in line with the data warehouse (Kimball versus Inmon).
- ETL, scheduling, metadata management, and governance should be aligned tightly with the data warehouse processes.
- The lowest level of detail is preferred; if summary data is also present in the data warehouse for the same subject areas, it should be left out.
- Historical data changes (commonly known as slowly changing dimensions) design principles should be enforced in the analytics base data even if the data warehouse doesn't have that feature for the subject areas of interest.

Performance Variables

Once the base analytics datamart is complete, attention shifts to performance variables. Performance variables are basically aggregate data fields that help the data mining algorithms identify common patterns where detailed data may not be much different between various types of scenarios. Since the pre-diction is on a package grain (meaning one record per package is assigned a probability of delay), performance variables factor in events before and after the package record. For example, each package has a truck assigned to it and this information may not reveal any interesting patterns discriminat-ing between on-time and delayed shipments, but if we actually look at where the truck was before it got assigned to the package, we may find out that a higher probability of delay mostly occurs when a truck starts its shift. Another example could be of a collection of packages that when put in together causes one of them to be delayed. This information is not available in the package subject area when we look at one package record at a time. So we have to build performance variables to capture the overall state of the package with a context, and sometimes that is where the hidden discriminatory pattern lies.

Chapter 11 explains some of the common techniques and methods of building performance variables, but this is also an art form. The more creative the performance variables, the better the chance for the data mining engine to find something interesting that has a higher discrimination value between good and bad. There are three types of performance variables:

1. Reporting variables
2. Third-party variables
3. Aggregate variables

Reporting Variables

Reporting variables are the most simple to build. Review the data warehouse reporting layer to see what kinds of reports are consumed by the business on the subject areas that are of interest for the analytics problem. In this case, we've listed the subject areas earlier in the chapter. Take a look at all the reports that use these subject areas and start documenting the variables created specifically for the business using the detailed data. Work with the business users and the subject matter experts (SMEs) to understand what those variables are (typically metrics), how they are computed, and, most importantly, how they are used to derive business decisions. This part of the summarized data may already be in some datamarts or implemented in the report program. The analytics datamart will have to reimplement these matric variables if they are not available as a simple extract. Sometimes it is better to reimplement them so that the analytics team can fully understand how they are being computed and get ideas for more interesting variables along the theme of these metrics.

Third-Party Variables

Third-party variables are aggregated pieces of information that may not be in the enterprise at all. This data should be pulled in directly into the analytics datamart since it may not serve any purpose for maintaining it in the data warehouse. Also, the acquisition of this data may actually cost money. Examples here would be demographics data, road closures, weather data, traffic data, or any other hazardous or perishable material packaging data that may come from typical customers who ship packages containing that material. A one-time data pull should be sufficient initially, and if the training model starts to rely on that data, then a permanent integration interface may be necessary. Most shipping companies rely on GPS and traffic data for their live rerouting and estimating delivery times, so this may not be that difficult to acquire.

It is preferred to store this data as an aggregate rather than a detail. For example, traffic updates may be available every few minutes and weather data may be available every few hours. We may have to build typical weather data and then build a scale of adverse weather from 1 to 5, and the same for traffic. If that data is not currently used, then historically delayed shipments will not have this data available. This is a challenge, because even though historical data is not available, it seems to be a useful set of variables and can certainly be made available to go live. If there is a way to estimate this accurately for historical shipments, then this is useful, otherwise it is not. The training data has to have this so the model can learn its impact on delayed shipments and then apply the weights when actually computing the delay probability on a real-time basis.

Aggregate Variables

This set of variables is where the art form comes into play. The aggregation of base analytic data results into variables that can influence the model significantly, and if they don't, more innovative variables can be created. The aggregation can be an on-time series, meaning roll-up of detail data along time and then assigning it to the package record. It can be sequence based, meaning the order of events, steps, or stages are rolled up to be assigned to each package record. They can be aggregated along frequencies of occurrence and along volumes (sums, averages). Aggregation can also be coded by introducing new code forms, for example, a shipment starting before 9 a.m. can be coded as "early" and a shipment starting at the employee shift end can be coded as "overtime." Another form of aggregation is looking for a specific pattern and setting a flag to be "Yes" or "No"—for example, if the traffic delays never went beyond 20% of the typical travel times, then a flag can be set to "No," indicating if traffic delays were encountered.

Model and Characteristics

The next component of the analytics datamart is the storage of the actual model and its characteristics. Building or training of an analytics model is an iterative process. We start by building one model based on a judgment call. Then we validate that model for its accuracy and false positives (i.e., declaring something to be bad that turns out to be good). Once the model starts to perform at a reasonable level or accuracy, it will move into production. Then, three months later, another test will be performed to see if the model is still performing at an acceptable level of accuracy. If not, it will go through a tuning exercise, which is very similar to the original model building exercise but a little more focused. This entire exercise has to be recorded in the analytics database. Once the team becomes comfortable with the model creation, validation, and tuning, previous generations of the model can be archived and do not need to be stored in the analytics datamart.

As explained in Chapter 4 on performance variables, the characteristics are variables that have shown a higher discriminatory value for identifying good versus bad data, therefore a performance variables data model within the analytics datamart may have hundreds or even thousands of variables, but a model may not use more than 20 or so. This model and characteristics component will only record the variables that have qualified to be characteristics. The models will be maintained as a generation and their proper version control naming will be recorded in the database. Once the training data has been prepared, it will be stored in this component. As the model going through the iterations keeps changing the characteristics it uses, this database has to track that as well as validation results. If the data mining software provides weights out of a trained model, then that has to be stored as well.

Model Execution, Audit, and Control

Once the model is ready and moved into production, as real-time or batch transactions come in that need to be predicted, the incoming characteristics and the predicted output both have to be recorded within this component. This is important from an audit standpoint as to which model was executed on a particular transaction and what probability was assigned. The need for this audit review can occur even after several months and years. Particularly in credit analytics, a loan that was approved based on a low probability of default may actually default after two years and a review may be required going back two years to see why it was approved.

Big Data, Hadoop, and Cloud Computing

When the idea for this book was originally conceived, Big Data and Hadoop were not the most popular themes on the tech circuit, although cloud computing was somewhat more prominent. Tablets and smartphones were still making waves, Facebook was still in its pre-IPO state, and social media, mainly through Twitter, had just demonstrated its revolutionary power in Middle Eastern politics with regime changes in Tunisia, Egypt, and Libya. Some of the reviewer feedback for this book suggested that these topics should be addressed in the context of the conceptual layout of analytics solutions so readers could make educated decisions whether they need to invest in newer skills or if they want to use their existing capabilities.

Throughout, this book has kept a technology-neutral posture speaking in general terms about databases, ETL and reporting tools, data mining and visualization tools, etc. The following topics will also get the same treatment, and their use in an overall analytics solution will be explained using the previous chapters as a foundation. Hadoop will be treated as yet another database or data processing technology designed for very large-volume data storage and analysis. The following three topics are presented as standalone material, each tying back into the overall analytics solution implementations presented in preceding chapters.

BIG DATA

If you are dealing with data beyond the capabilities of your existing infrastructure, you are dealing with *Big Data*; essentially, too much data is Big Data. Another popular definition that is somewhat fading away is that unstructured or newer forms of data that didn't exist until recently are considered Big Data (e.g., Facebook "likes," Twitter "tweets," or smart-energy meter readings). When the term *unstructured* was used, this included images, PDFs, document files, music files, and movie files, which were always digitally stored but hardly ever analyzed. However, the best definition for Big Data comes from Gartner Inc. (2012): *Big Data*, in general, is defined as high *volume*, *velocity*, and *variety* (the three V's) information assets that demand cost-effective, innovative forms of information processing for enhanced insight and decision making.

CONTENTS

This definition is where consensus is developing, and I will use it as a basis for explaining Big Data within analytics solutions. It has to be understood that without a sufficient amount of data, analytics solutions may not deliver the expected return on investment. Insufficient data volume limits the successful training of an analytics model, or even if it is successful, its performance is not ready for business decisions. So when does sufficient data become Big Data? Big Data refers to newer forms of data that we can now tackle in unprecedented sizes, shapes, and volumes. However, data is data, and it has to be analyzed, performance variables identified, models built, decision strategies designed, and business decisions made, tuned and monitored. Big Data also has to go through the same motions to deliver analytical value. In the Information Continuum discussion (see Chapter 2) the hierarchy of data utilization starts at raw data and, as our understanding and comfort of data increases, we move up that hierarchy extracting higher value from data. Big Data will also have to go through the same process, where it will start at raw data and move up as we use a different set of tools capable of handling it.

The definition of Big Data uses three characteristics: velocity, variety, and volume. Let's look at each of them in detail. Whether data requires all three or just one to be qualified as Big Data is a debate that will take a few years to settle as this area matures. For now, we will assume that any one of the three characteristics present qualifies a problem as Big Data.

Velocity

Velocity refers to the speed with which data is generated. Data generated from user interaction is limited by the number of users and the transactions being performed. Since Big Data has a notion of very high velocity, we can argue that traditional business and commerce-related transactions cannot be easily qualified as Big Data; besides, traditional database systems have been handling those interactions reasonably well and their volume is unlikely to increase several-fold overnight. So what are some situations where data gets generated at a very high velocity? Most situations involve machines and devices generating data. However, there are some unique situations in web searches, very large social media sites, or gaming platforms on the web where hundreds of millions of users can simultaneously be generating a lot of data from their normal activities in those environments. Wireless devices constantly communicating with cellular towers almost every second, emission sensors on-board an automobile detecting carbon contents every few milliseconds, oceanographic sensors detecting tsunamis, weather sensors recording moisture, wind, and temperature, etc., are all examples of velocity where the speed of recorded data is overwhelming for the traditional hardware and relational database systems. Not only that, a data collection and analysis problem in high-velocity situations can involve thousands of sensors all recording and reporting data with subsecond frequencies, which creates a communication, storage, and processing problem all at once for traditional computing infrastructure.

Variety

Variety in Big Data deals with the variation in the records getting generated, meaning how many different kinds of data are being recorded. The examples used to explain velocity deal with limited variety, because no matter how many sensors are reporting data or how frequently they report it, if the information set just contains six to eight fields and the variation of data in those fields is also very limited through coded values or well-defined ranges, then the variety of data is low. On the other hand, user activity on an interactive cable-TV box or smart-TV will result in generating all sorts of records from channel viewing, to channel skipping, to program details, to advertisement durations, to DVR, to on-demand viewing, to premium channels, etc. If millions of users are simultaneously interacting, then the data has both velocity and variety. Popular social media platforms also have both velocity and variety since millions of users are interacting and their interactions generate a wide variety of different data points. Variety deals with both different layouts of records as well as variation in possible values in the fields within the records.

Volume

Volume deals with the size of the data required for analysis. It can include longer histories, such as weather sensor data recorded and analyzed over several years to predict weather system movements. Volume can also refer to a large number of users performing the same activity (clicking on a breaking story), and therefore the specific scenario that deals with that situation is overwhelmed with the volume—although depending on what is being analyzed, it may just be a problem of velocity. Volume deals with both the storage and processing of large data sets. In the absence of Big Data technology and toolsets, analysts working to build predictive models could not use this kind of volume (hundreds of terabytes), and therefore always used a representative sample. However, with a wide variety of data mining algorithms and cheaper hardware resources, they can now tackle the problems without bringing its volume down to a few gigabytes through sampling and losing information in the process. The NoSQL initiative was in fact in response to this exploding volume of data which would have cost millions in hardware infrastructure and still the relational database engines would not have been able to handle this kind of data volume. Hadoop is just one implementation of that NoSQL movement that has been very successful at tackling unprecedented problems of volume. Here is a reference for more detail on the NoSQL initiative http://nosql-database.org/.

Big Data Implementation Challenge

Big Data has challenges both in the operational environment and in analytical environments. In operational environments, Big Data challenges can deal with overwhelming traffic on a website because of an event or a cyberattack, and dealing with that requires newer forms of tools and technologies. Diagnostic applications on-board aircrafts and heavy machinery deal with a large volume of sensory input, and based on that they have to take a course of action. However, we will limit ourselves to a Big Data implementation challenge only within the analytical space where historical data is essential in identifying patterns that can be used for proactive decision-making strategies. Two excellent articles provide more depth to this topic:

- "What Big Data Is Really About" (Madsen, 2013)
- "What's Your Strategic Intent for Big Data?" (Davenport, 2013)

The technology vendors dealing in NoSQL databases, in-memory systems or database appliances typically position their Big Data solutions as an alternate to traditional data warehousing. The premise that drives this perspective is that savvy business users want all of their data all the time and in one place so they can get to it any time and anyway they want to without going through months of development through various marts and summary layers. If that is in fact a challenge posed by the business to the data warehousing and IT teams, the proposition that buy a big enough appliance or big enough data storage and processing infrastructure and eliminate all the ETL and aggregations and marts and summaries, etc. may actually make sense. However, the Information Continuum prohibits or creates a barrier to this approach. While it is technologically possible to build a very large Hadoop cluster and dump all the data from all the internal and external systems in there, the understanding of the inter-relationships of the data and exposing all that data meaningfully to business users would be a challenge. Users can get lost in such a large universe of data fields and records and their inter-relationships and dependencies or they can inter-mix data incorrectly without realizing. The Information Continuum represents a natural evolution where organizational appetite for data and insights as well as the understanding also evolves accordingly.

Controlling the Size

The first thing to identify is whether the problem domain under consideration poses all three problems of velocity, variety, and volume. If the problem primarily deals with machine sensor–type data that sends readings every few milliseconds, there is a way to eliminate the velocity and volume parameters from the equation by only recording significant shifts in the readings. As long as the sensor sends the same exact information, there may be an option to ignore it. However, careful analysis of the problem domain is required to see whether the readings before the significant shift actually demonstrate a pattern. Keeping all the detailed readings can come in handy, but the effort and cost required to manage that should be justified against stated objectives.

Similarly, if the variety in data is overwhelming, we want to make sure that each data type has enough volume represented in the analysis. If the volume is skewed toward a handful of record types, then we may need to apply some principles from statistics to bring the volume across the varieties to representative sizes. The key to managing the size is to see if one or two of the three characteristics (the three V's) can be eliminated to manage cost and scope of the problem across the data size, problem statement, analytics models, ensuing decisions, and expected results.

Applying the Information Continuum

Once the three characteristics are analyzed and the problem parameters are understood, analyzing the data starts with the Information Continuum i.e., search/lookup, then counts, summaries, reporting and eventually analytics. Without this process up to this point it is difficult to fully understand the data and therefore there is a limited ability to get value out of this data. Remember, this is data that business has never analyzed before, and therefore they would need this capability to get comfortable and start formulating what they want to do with it. From basic reporting all the way to analytics modeling in the Big Data problem space, all the stages of Information Continuum may not be needed as a specific implementation layer but the understanding is required. Once the comfort is there with the Big Data and a formulated problem statement, you can jump directly into the analytics model. The application of analytics techniques, performance variables, and all the other components of the analytics solution are as applicable to Big Data as they are to traditional types of structured data.

HADOOP

Creation of Hadoop can be traced back to 2005 as an initiative at Yahoo by Doug Cutting driven or inspired by Google's MapReduce technology (Vance, 2009). By 2009, Hadoop dominated web searches and large websites' internal workings to organize, index, and search troves of data and service ads at companies like Yahoo, Google, and Facebook. Hadoop is a file system capable of storing and processing an unprecedented amount of data presented in any file format. It uses a technology called MapReduce to search, access, and manipulate that data while running on cheaper interconnected computing machines (even old PCs).

Hadoop Technology Stack

The best representation of Hadoop's technology stack that I found relevant to the introductory level of detail presented here comes from Michael Walker (2012). Figure 11.1 shows where a Hadoop technology stack fits into the

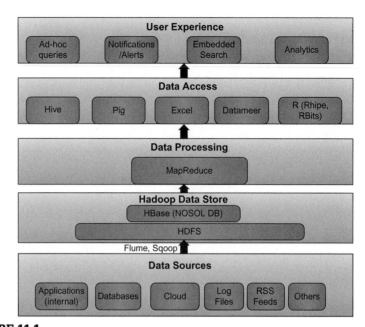

FIGURE 11.1

Hadoop technology stack. *Source: Walker (2012).*

overall data processing solution and what kind of tools are available in each layer of the stack for specific functions.

Data Sources

Data sources refer to data, or rather Big Data, that needs to be brought into the Hadoop file system (HDFS). Various tools and software packages exist that allow moving data from conventional storage like UNIX file systems or relational database systems or even from logs or various other forms of storage into Hadoop. Hadoop takes the incoming data and loads it into its own file system while structuring the data across the various clusters (groups of computers or nodes to run Hadoop) distributing the incoming data across the nodes in the cluster.

Hadoop Data Store

The Hadoop data store has HDFS as the file system and a catalog that tracks where data has been stored. The files from the data sources (e.g., logs, PDFs, images, etc.) do not retain that native structure, rather they are converted into the HDFS format. Unlike Windows File Explorer, which shows all the files on the Windows file system like documents, spreadsheets, etc., you cannot open the HDFS and look at the original files easily.

Data Processing

Once the data is within the HDFS, the only way to access it is through the MapReduce command interface. The MapReduce command interface allows for entire processing logic to be written in MapReduce. However, MapReduce programming is not trivial since it requires breaking down the processing logic into a parallel rather than sequential programming code. Breaking down a business problem in the form of Map and Reduce functions is quite a programming challenge.

Data Access

to address this data processing challenge of MapReduce, which is not rich enough like a DotNet programming environment, an entire data access layer has been built over time. This data access layer consists of a wide variety of open-source and proprietary tools and libraries developed for various programming needs. If MapReduce was like Assembly Language programming, the data access layer is more like C/C++, SQL, and Java type of programming, which is more business friendly.

User Applications (User Experience)

The user application or, as Michael Walker puts it, user experience layer is like the application layer where complex business logic is combined to deliver value. This is the fastest growing space within the Hadoop technology stack where libraries, tools, software packages, and suites are starting to become available in a wide variety for specific business applications.

Hadoop Solution Architecture

If a corporate organization wants to adopt Hadoop for some of their complex business requirements dealing with Big Data, the deployment architecture of Hadoop is still very complicated. However, the good news comes from a wide variety of software packages that are now Hadoop compliant. Almost all the big technology vendors have adopted Hadoop as an answer to very large data-centric problems primarily for two reasons:

- Hadoop is extremely scalable.
- Hadoop implementation in terms of hardware cost is very low (Davenport, 2013).

Organizations should look for vendors that provide the data access or user application layer tools that are Hadoop compliant instead of writing their own until the technology matures to a point where skill and support is readily available. Deploying Hadoop alone will be a challenge for typical internal IT staff, as its deployment has a lot of moving parts and standardized documentation and structured methodologies are still developing.

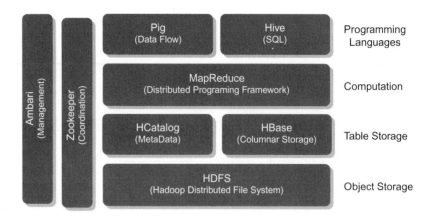

FIGURE 11.2

Hadoop technical architecture components. *Source: Walker (2012).*

Figure 11.2 from Michael Walker's blog is a simplistic component diagram that shows the relevant layers needed to make use of Hadoop for a business application. On the left side it shows two additional layers that allow for managing the entire environment. Still, archive and retrieval, failover, disaster recovery, data governance, and information security controls are in their infancy compared with established database systems like Oracle or Teradata. Hadoop can be used in operational and transactional environments where large volumes of data need to be analyzed as quickly as it is created to look for certain thresholds that are being broken or specific data pattern needs to be identified. A trigger-based approach is typically used and a massive amount of sensor or web log for user activity data passes through Hadoop as it looks for specific predefined triggers of very high readings from sensors or specific user activity. These are caught and reported in real time to users or other applications, since Hadoop deals with all three V's at the same time. This is the operational perspective of Hadoop and we will not delve further into its implementation.

The use of Hadoop to solve Big Data analytical problems has two variations:

1. Hadoop acting as an ETL layer to process Big Data and loading into a RDBMS-based traditional analytics datamart (Chapter 10 describes analytics datamarts in great detail).
2. Hadoop acting as the data mining engine processing data to build a model.

Hadoop as an ETL Engine

Figure 11.3 shows how Hadoop would fit into an analytics solution. The idea in this approach is to aggregate or build performance variables from the Big

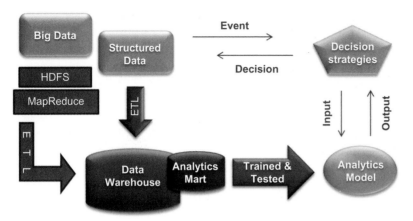

FIGURE 11.3

Hadoop as an ETL engine.

Data using Hadoop, while traditional data that needs to be intermixed takes a conventional ETL route. Once the Big Data is reduced to more manageable sizes (by eliminating one or more V's from its characteristics), it can be treated as conventional structured data that can be stored and processed in a relational database system. In this scenario, Hadoop is acting like a lean and efficient ETL component. Leading ETL vendors have added Hadoop support in their suite of tools to do this from within one data integration suite (Thoo et al., 2012).

Figure 11.3 is a variation of the same diagram as we saw in chapter 9 on Analytics Implementation methodology (Figure 9.3) but here, Hadoop is shown as an added ETL layer for Big Data. This advanced form of ETL is required to convert unstructured data like tweets, videos, phone calls, machine sensor logs, etc. into structured data. This is important because unstructured data may not be very useful to be used in analytics models unless the knowledge from that data is integrated with other structured business and operations data in order to build better performance variables and analytics models. Therefore, treatment of unstructured data should be considered a separate layer of specialized ETL (on steroids) that can read and decipher the digital structure behind unstructured data and get value out of it in structured form and feed into a more conventional layer of ETL using Hadoop that is then able to integrate that with other Big Data feeds. If after applying Hadoop to unstructured data, the variables and structured data that we extract are manageable in size, then they can be run through a conventional ETL layer as well following traditional data warehousing methods.

Hadoop as an Analytical Engine

The other option for Hadoop within an analytics solution is where the entire data mining algorithm is implemented within the Hadoop programming

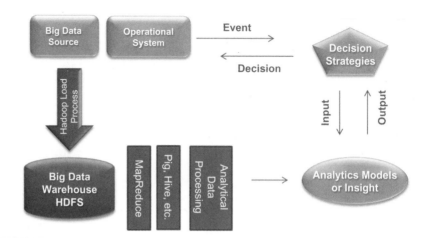

FIGURE 11.4

Hadoop as an analytical engine.

environment and it acts as the data mining engine shown in Figure 11.4. This is used when there is no option of reducing, aggregating, or sampling the data to eliminate the V's from Big Data characteristics. This becomes quite complex as the performance variables, the heart of innovation within analytics modeling, cannot be easily added to the data set without creating an additional storage layer. However, various problems actually require running the entire Big Data set looking for specific trends or performing fuzzy search or correlations between variables through Hadoop and its data access programs.

Outside of analytics—as defined in the first chapter—Hadoop can be used as generic data processing engine that can search and look for patterns and correlations almost like a reporting tool of sorts. It should be treated as a commodity data processing layer that can be applied to any overwhelming data size problem.

Big Data and Hadoop—Working Example

Let's take a Big Data problem from the retail shopping sector. A large brick and mortar retailer with thousands of square footage in floor space wants to understand the following three things.

1. How are the floor placement of promotions and products attracting people i.e., how many people stop and how many ignore and pass?
2. How many people actually enter the store broken down by entrance and time of day and how does that correlate with sales?
3. What is the probability that a customer stopping at a display will buy the product?

In order to answer these questions, it is evident that a new form of data is required. The traditional structured data available with retailers such as point of sale, inventory, merchandising, and marketing data cannot answer these questions. Therefore an additional source of data is required to answer these questions. As most stores are fitted with security cameras and the digital surveillance technology has improved dramatically, we will assume that the security cameras are available across the store and they do have digital footage from the stores 24/7.

Dozens of video files are available from security cameras on a daily basis and for a large retail chain the number could be in thousands. The video files are typically voluminous, so we are certainly looking at the "V" of volume to run this analysis. It is also certain that the "V" of variety is not applicable here and that velocity will be a challenge if we decide to do this analysis in real time as videos are streaming. So certainly, this is a Big Data problem.

Specialized code will be required to analyze the videos and separate people, entrance and floor layout. Some kind of logical grid of the floor plan will be needed and then the displays will have to be marked in that logical construct. In the first step, we will use Hadoop as an ETL engine and process all the video files.

For Question 1: The output should provide the required metrics by each display as the video image processing logic (implemented using Hadoop technology stack described above) should identify the display, count the people passing, and count the people stopping. This is a massive computing challenge in storage and CPU and Hadoop is exactly designed to handle this. The structured output data should be recorded in the data warehouse

For Question 2: This one combines the data extracted from the videos using the Hadoop implementation with some structured data available within the data warehouse system as Point-of-Sale(PoS) systems track the sales transactions and have a timestamp on them. So the same logical grid of the floor plan will be used to identify entrance and video imaging program (implemented in Hadoop technology stack) will count people entering and leaving. This output will be recorded into the data warehouse that already has the information on PoS. With the two datasets in structured form now recorded in the data warehouse, all sorts of additional questions can be asked intermixing the data

For Question 3: This one is an extremely complicated question because it requires identifying the customers with a unique ID so they can be tracked passing through displays as well as when they bought something. We will first limit our solution to the video footage of shoppers who stopped at a display. At the time of purchase, we will need structured data to link their loyalty card or another form of ID we may already have on file with this video file ID. The two streaming videos have to be combined, but we do not know which ones might have the same customer in them although we can limit ourselves to the same store footage and within the 2–3 hour window of the video showing someone entering. This completes one part of the solution where we are able to track the ones who stopped at a display and also bought something (this makes up our "1" records for predictive model training). The other part of the data (the "0" records) who stopped at a display but did not buy will be the remaining population in the video footage with a unique ID. Now that we have two population sets, we can start to work on performance variables and as described in Chapter 4 on performance variables, there is a lot of freedom in looking for variables from the videos, like clothing, individual or family, young or old, other shopping bags in hand, food in hand, on phone or not, etc. are all variables that can be extracted from the video files. The image processing technology has dramatically improved in the past few years so do not be overwhelmed by the massive video image processing required here. Once the variables are identified, the model needs to be trained and then if the predictions are needed in real time for the ones who may not buy so a sales associate can attend to them then we are dealing with the "V" of velocity as well. This type of problem requires the entire solution to be coded in the Hadoop cluster because even if we take the performance variables out as structured data and use conventional data mining algorithm to build the model, for real-time use, the video has to be constantly processed and run through the model. Anything outside of Hadoop environment may create a scalability bottleneck.

CLOUD COMPUTING (FOR ANALYTICS)

Cloud computing has taken the information technology industry by storm. The idea is that managing massive technology infrastructure and environments is extremely expensive and ensuring reliability, security, and availability of that infrastructure requires specialized technology and skill. If all an organization wants to do is run a few essential business applications for finance, HR, and operations, why do they have to incur the significant upfront and operational costs of managing the infrastructure? The idea of letting a third party take care of the IT portion of a business makes a lot of sense, and CFOs and other business executives are in agreement. However, using cloud computing for business applications means data will be outside the walls of the organization, and there is a certain degree of discomfort with that, thereby limiting the use of cloud computing mostly for nonessential business functions.

Disintegration in Cloud Computing

The biggest challenge for data warehousing and analytics because of the introduction of cloud-based application systems is data integration. Over the last three decades we finally accomplished a very daunting and complex challenge of data integration across the enterprise. Now again as some systems are moved to cloud computing, data integration will again become a challenge. The tools in cloud computing to handle complex data integration, particularly semantic mapping and data quality, integrity, and scalability, have not yet been adequately addressed.

When an application system is moved to cloud computing in a SaaS (software as a service) environment, the cost savings come from adopting the SaaS software functionality as it is offered. There is limited opportunity for customization because then the economies of scale for the SaaS go away. This also means that how the SaaS vendor defines data fields, their valid values, their interrelationships, etc., has to be accepted. However, the data warehouse and the analytics solutions are probably using the older definitions from the legacy system that is being replaced, and therefore the entire data movement process from the legacy system needs to be reviewed in light of the new data feeds that will have to come from the SaaS application. This can be quite a challenge, and additional cost of reimplementation has to be factored in to the overall SaaS project's budget.

Analytics in Cloud Computing

Is cloud computing suited for analytics solutions? There are some areas of analytics solutions where cloud computing makes a lot of sense, particularly when you need cheap storage and cheap processing power for very large data processing tasks. However, the risk perception of data loss and data breach will probably keep the complete analytics solutions from using cloud computing. The niche players in cloud computing that offer analytics solutions are actually very attractive and should be looked at. There are several start-ups that offer price optimization cloud-based solutions or smart-meter energy usage forecasting.

This area of analytics has a lot of potential because entrepreneurs having depth in any industry can invent newer Big Data solutions, develop them once in the cloud, and offer them as SaaS. They can receive data on a regular basis, process it, build and tune client-specific models, and then run their transactions through them. The data input and output mechanism is actually well structured since the SaaS solution knows all the various performance variables and characteristics it needs. With the infrastructure in place, the SaaS vendors can try newer value propositions out of the same data domain and keep inventing and extracting more value out of data, while the client can use that output innovatively into decision strategies and maintain a competitive edge even when they are using the same SaaS as their competitor.

Conclusion

In the Introduction, the four objectives for writing this book were listed: simplification, commoditization, democratization, and innovation. This chapter sums up the entire book's contents in support of meeting those objectives.

OBJECTIVE 1: SIMPLIFICATION

The simplification objective is achieved at several levels, as follows.

Simplified Definition

When a technology or concept is overhyped in the market, it becomes very difficult for new entrants to filter out the marketing buzz. In case the of analytics, the mathematics, statistics, and machine learning theories (data mining) are already very complex and intimidating, let alone understanding their proper application and implementation in a commercial organization. Therefore, Chapter 1 defines an analytics solution and puts it in a business context. As stated numerous times, the purpose is to simplify the definitions so they can be adopted and implemented, therefore the definitions have been kept very simple. Once an organization successfully implements their first project, demonstrates value, and the IT and business sides become confident and excited, a more sophisticated deep-dive into the algorithms and techniques will evolve naturally.

Demystifying Analytics Techniques

Chapter 1 uses just four techniques as an arbitrary way to limit the complexity of available analytics techniques. Also, the techniques chosen and explained are simpler to understand compared with, let's say, social network analysis or association rules, which are difficult to apply in a direct and relevant business context. Being introductory material regarding analytics, the techniques chosen were explained further through examples in Chapter 3 on using analytics. What readers will find is predictive analytics is the most powerful of all the available techniques, which has been explained in great lengths in terms of concept, examples, implementation details, and even planning and design suggestions throughout Chapters 4 and 5, as well as Chapters 7, 9, and 10.

Simplifying Implementation Details

Chapters 9 and 10 cover the IT and implementation in quite a bit of detail, emphasizing the tasks, deliverables, roles, responsibilities, and skills needed to put a team together that can deliver analytics solutions repeatedly and consistently. Chapter 11 covers Big Data, Hadoop, and cloud computing to complete the technology landscape surrounding analytics solutions. Similar to the analytics definition, these concepts are also in their infancy from a maturity and wide adoption perspective, and therefore it is important to filter material content from marketing buzz.

OBJECTIVE 2: COMMODITIZATION

Commoditization refers to using readily available tools and existing infrastructure that are usually available within an organization, such as ETL tools, a data warehouse system, etc. Chapter 5 shows how decision strategies can be implemented using an existing ETL toolset that includes messaging or service-oriented integration platforms. Chapter 9 details the implementation methodology and shows how an existing data warehouse environment should be used, and Chapter 6 introduces and emphasizes this same point. Chapter 2 on data utilization presents a comprehensive tool that organizations can use to see where they stand on this Information Continuum. Once they place themselves, they can see the existing toolsets, skills, and infrastructure available to them and how to build from there to get to analytics solutions dealing with models and decision strategies.

OBJECTIVE 3: DEMOCRATIZATION

Democratization refers to making analytics available to a wider audience compared with a select few. Chapter 7 takes an example and emphasizes the business case and demonstrated value allowing for a wider audience to take notice. Democratization has three different perspectives. The first is making analytics solutions available to an entire organization and not just a small set of specialized areas. This point is made repeatedly in several chapters, especially in Chapter 5 on decision strategies. The second perspective applies to industries that have traditionally not adopted analytics as a tool for competitive advantage. Industries like shipping and logistics, education, the public sector, etc. are not well known for their use of analytics. They can now adopt the simplified and commoditized analytics solutions, as the initial solutions will be cost effective, and once value is demonstrated, skills are developed, and a culture emerges, it becomes easier to become an analytics-driven organization. Chapter 3 on using analytics provides numerous examples from a wide cross-section of industries to help readers see a pattern within those examples and then start to look for opportunities within their industry. The third perspective is focused on midsize organizations that have the same problem as a large corporation but may be smaller in scale. A regional bank, for example, that has a small credit derivatives portfolio cannot hire a team of experienced traders or invest in high-end analytical tools to run an efficient derivatives trading desk. With a simplified and commoditized solution, they now have an opportunity to apply portfolio optimization and forecasting techniques to improve their returns and use predictive modeling to manage their risk more effectively.

OBJECTIVE 4: INNOVATION

Does analytics allow a business to become innovative? The approach taken in Chapter 3 uses examples across various functions and industries where analytics can provide a competitive edge. The real case for innovation comes through in Chapter 8 on requirements and scope. Identification of a problem or realization of an opportunity is at the heart of becoming an innovative business using analytics. The analytics models aside, the decision strategies explained in great detail in Chapter 5 show how output from an analytics model can be used to innovate within business operations through small incremental gains. If eventually an organization is able to instill a thinking culture along these lines, then there is no end to human creativity yielding value for the business.

A decision strategy–driven culture with output coming from models that are always updated based on changing business environments starts to make middle managers and then even line workers get into the habit of constantly innovating and evolving their decision strategies for maximizing value. They would almost start to become innovative business executers by habit (Duhigg, 2012).

References

ADP Corporation. (2012). *ADP history*. Retrieved from ADP: <http://www.adp.com/careers/uscareers/who-is-adp/history.aspx>. Accessed December, 2012.

Bátiz-Lazo, B., Maixé-Altés, C., & Thomes, P. (2010). *Technological innovation in retail finance: International historical perspectives*. London: Routledge.

Chatterjee, S., & Hadi, A. S. (2006). *Regression analysis by example*. New York: Wiley-Interscience.

Davenport, T. H. (2010). Competing on talent analytics. *Harvard Business Review*.

Davenport, T. H. (2012). Data scientist: the sexiest job of the 21st century. *Harvard Business Review*.

Davenport, T. H. (2013). What's your strategic intent for big data? *CIO Journal*.

Drucker, P. (2002). The discipline of innovation. *Harvard Business Review*.

Duhigg, C. (2012). *The power of habit: Why we do what we do in life and business*. New York: Random House.

Eckerson, W. (2010). *Performance dashboards: Measuring, monitoring, and managing your business*. New York: Wiley.

Eckerson, W. (2012). *Secrets of analytical leaders: Insights from information insiders*. Technics Publications.

Equifax. (2012). Retrieved from Equifax: <www.equifax.com>. Accessed December, 2012.

Experian. (2012). Retrieved from Experian Inc.: <www.experian.com>. Accessed December, 2012.

FICO. (2012). *FICO score*. Retrieved from FICO Corp.: <www.fico.com>. Accessed December, 2012.

Financial Crisis Inquiry Commission. (2011). *The financial crisis inquiry report*. Washington, DC: US Government Printing Office.

Gartner Inc. (2012). *Big data*. Retrieved from Gartner.com: <http://www.gartner.com/it-glossary/big-data/>. Accessed December, 2012.

Graunt, J. (1662). *Natural and political observations made upon the bills of mortality*. London.

Haider, S. (2012). *Dr. Sajjad Haider*. Retrieved from IBA: <http://sajjadhaider.iba.edu.pk/teaching.html>. Accessed December, 2012.

Han, J. (2012). *Jiawei Han*. Retrieved from University of illinois Urbana-Champaign: <http://www.cs.uiuc.edu/~hanj/>. Accessed December, 2012.

Han, J., & Kamber, M. (2011). *Data mining: Concepts and techniques* (3rd ed.). San Francisco: Morgan Kaufmann.

Hohpe, G., & Woolf, B. (2003). *Enterprise integration patterns: Designing, building, and deploying messaging solutions*. Reading, MA: Addison-Wesley Professional. Retrieved from Wikipedia: <http://en.wikipedia.org/wiki/Enterprise_service_bus>. Accessed December, 2012.

Ifrah, G. (2002). *The universal history of computing: From the abacus to the quantum computer.* New York: Wiley.

Inmon, W. H. (1992). *Building the data warehouse.* Wiley.

Inmon, W. H. (2005). *Building the data warehouse.* Wiley.

Kaplan, R., & Norton, D. (1992). The balanced scorecard—measures that drive performance. *Harvard Business Review.*

Kimball, R. (2002). *The data warehouse toolkit: The complete guide to dimensional modeling.* New York: Wiley.

Kimball, R., Ross, M., Thornthwaite, W., Mundy, J., & Becker, B. (2008). *The data warehouse lifecycle toolkit.* New York: Wiley.

Lewin, R. (2001). *Ultra goes to war.* New York: Penguin Press.

Linoff, G., & Berry, M. (2011). *Data mining techniques: For marketing, sales, and customer relationship management.* New York: Wiley.

Machine Learning Group. (2012). WEKA. New Zealand. Retrieved from <http://www.cs.waikato.ac.nz/ml/index.html>. Accessed December, 2012.

Madsen, M. (2013). What big data is really about. *TDWI–BI This Week.*

Markillie, P. (2012). A third industrial revolution. *The Economist.*

Mathworks. (2012). *Matlab.* Retrieved from Mathworks: <http://www.mathworks.com/products/matlab/>. Accessed December, 2012.

May, T. (2009). *The new know: Innovation powered by analytics.* New York: Wiley.

Mearian, L. (2007). A zettabyte by 2010: Corporate data grows fiftyfold in three years. *Computerworld.*

Mendenhall, W., & Sincich, T. (2011). *A second course in statistics: Regression analysis.* Pearson.

Merriam-Webster. (2012). *Merriam webster.* Retrieved from <http://www.merriam-webster.com/>. Accessed December, 2012.

Murty, K. G. (2003). *Optimization models for decision making.* Retrieved from University of Michigan: <http://www-personal.engin.umich.edu/murty/>. Accessed December, 2012.

Murty, K. G. (2009). *Optimization for decision making: Linear and quadratic models.* New York: Springer.

Nocedal, J., & Wright, S. (2006). *Numerical optimization.* New York: Springer.

Oxford Dictionary. (2012). *Oxford dictionary.* Retrieved from Oxford Dictionary: <http://oxforddictionaries.com/>. Accessed December, 2012.

Pondera Consulting. (2012). Fraud investigation dashboard. *Fraud Dashboard.* <www.pondera-consulting.com>. Accessed December, 2012.

R Foundation. (2012). *R project.* Retrieved from The R Project for Statistical Computing: <http://www.r-project.org/>. Accessed December, 2012.

Reason, J. (1990). *Human error.* Cambridge University Press.

Redman, T. C. (2008). *Data driven: Profiting from your most important business asset.* Cambridge, MA: Harvard Business Press.

Russell, S., & Norvig, P. (2009). *Artificial intelligence: A modern approach.* Englewood Cliffs, NJ: Prentice Hall.

SAS Corporation. (2012). *SAS.* Retrieved from <www.sas.com>. Accessed December, 2012.

Scott, C. A. (1907). *Cartesian plane geometry.* J.M.Dent. Retrieved from wikipedia.org.

SPSS. (2012). *SPSS.* Retrieved from IBM: <http://www-01.ibm.com/software/analytics/spss/>. Accessed December, 2012.

Steiner, C. (2012). *Automate this: How algorithms came to rule our world*. New York: Portfolio / Penguin.

Taleb, N. N. (2008). *Fooled by randomness: The hidden role of chance in life and in the markets*. New York: Random House.

Taleb, N. N. (2010). *The black swan: The impact of the highly improbable* (2nd ed.). New York: Random House.

Thibodeau, P. (2012). Big data brings big academic opportunities. *Computerworld*.

Thoo, E., Friedman, T., & Beyer, M. A. (2012). *Magic quadrant for data integration tools*. Gartner.

TransUnion. (2012). Retrieved from TransUnion: <www.transunion.com>. Accessed December, 2012.

Vance, A. (2009). Hadoop, a free software program, finds uses beyond search. *New York Times*. Retrieved from <http://www.nytimes.com/2009/03/17/technology/business-computing/17cloud.html?_r=0>. Accessed December, 2012.

Walker, J. (2012). Meet the new boss: big data. *The Wall Street Journal*.

Walker, M. (2012). *Hadoop technology stack*. Retrieved from Analytic Bridge: <http://www.analyticbridge.com/profiles/blogs/hadoop-technology-stack>. Accessed December, 2012.

Warwick, K. (2011). *Artificial intelligence: The basics*. London: Routledge.

Watson, H., Gerard, J., Gonzalez, L., Haywood, M., & Fenton, D. (1999). Data warehousing failures: case studies and findings. *Journal of Data Warehousing*.

Watt, D. (2003). *E-business Implementation: A guide to web services, EAI, BPI, e-commerce, content management, portals, and supporting technologies*. Boston: Butterworth-Heinemann. Retrieved from Wikipedia: <http://en.wikipedia.org/wiki/Enterprise_application_integration>. Accessed December, 2012.

Witten, I., Frank, E., & Hall, M. A. (2011). *Data mining: Practical machine learning tools and techniques*. San Francisco: Morgan Kaufmann.

Index

Note: Page numbers followed by "*f*" and "*t*" refer to figures and tables, respectively.